NATURAL DISASTER RESEARCH, PREDICTION AND MITIGATION

EMERGENCY MANAGEMENT

AN OVERVIEW AND ISSUES FOR CONGRESS

Natural Disaster Research, Prediction and Mitigation

Additional books in this series can be found on Nova's website under the Series tab.

Additional e-books in this series can be found on Nova's website under the eBooks tab.

NATURAL DISASTER RESEARCH, PREDICTION AND MITIGATION

EMERGENCY MANAGEMENT

AN OVERVIEW AND ISSUES FOR CONGRESS

STEPHANIE PADBURY

snova
New York

Library of Congress Cataloging-in-Publication Data

ISBN: 978-1-53614-101-6

Published by Nova Science Publishers, Inc. † New York

CONTENTS

PREFACE

After a flood, people are often uncertain if their eligibility for federal disaster assistance is linked to any way to whether or not they have flood insurance. The first two chapters in this book provide an overview of the assistance available to individuals and households following a flood and provides links to more comprehensive guidance on both flood insurance and disaster assistance. Introductory information on key components of the National Flood Insurance Program (NFIP), ranging from floodplain mapping to the standard flood insurance forms are provided. The next chapter examines the First Responder Authority (FirstNet), a federal agency created by Congress to help public safety agencies to communicate during incidents (during the events of September 11, 2011, first responders could not communicate with each other. Some radios did not work while other radio channels were overloaded by a large number of responders trying to communicate). The subsequent chapter briefly describes recent education-related developments in response to numerous disasters in 2017. It then describes education-related federal assistance provided by FEMA in the aftermath of disasters generally. This is followed by an examination of education-related disaster recovery and issues related to elementary and secondary education, postsecondary education and various ED-administered programs typically serving adult learners. The fifth chapter provides a short summary of the types of individual assistance programs administered by FEMA (Federal Emergency Management Agency) following a disaster. It also provides a summary of the criteria FEMA uses in determining which individual assistance programs may be made available to impacted areas following a major disaster declaration, and discusses a proposed rule to change these criteria. The last chapter provides an overview of the United States Fire Administration (USFA), whose objective is to significantly reduce the nation's loss of life from fire, while also achieving a reduction in property loss and nonfatal injury due to fire.

In: Emergency Management
Editor: Stephanie Padbury

ISBN: 978-1-53614-101-6

Chapter 1

INTRODUCTION TO THE NATIONAL FLOOD INSURANCE PROGRAM (NFIP)*

***Diane P. Horn*[1] *and Jared T. Brown*[2]**
[1]Flood Insurance and Emergency Management
[2]Emergency Management and Homeland Security Policy

ABSTRACT

The NFIP was established by the National Flood Insurance Act of 1968 (NFIA, 42 U.S.C. §4001 et seq.), and was most recently reauthorized to July 31, 2018, through a series of short-term re-authorizations. The general purpose of the NFIP is both to offer primary flood insurance to properties with significant flood risk, and to reduce flood risk through the adoption of floodplain management standards. Communities volunteer to participate in the NFIP in order to have access to federal flood insurance, and in return are required to adopt minimum standards.

FEMA manages a process, called Risk MAP, to produce Flood Insurance Rate Maps (FIRMs). Depicted on FIRMs are Special Flood Hazard Areas (SFHAs), which is the area exposed to a 1% or greater risk

* This is an edited, augmented and reformatted version of a Congressional Research Service report, R44593, dated April 30, 2018.

of annual flooding. FIRMs vary in age across the country, and are updated on a prioritized basis. The Risk MAP process provides extensive outreach and appeal opportunities for communities. Updating a community's FIRMs can take three to five years or more. Participating communities must adopt a flood map and enact minimum floodplain standards to regulate development in the SFHA. FEMA encourages communities to enhance their floodplain standards by offering reduced premium rates through the Community Rating System (CRS). FEMA also manages a Flood Mitigation Assistance (FMA) grant program using NFIP revenues to further reduce comprehensive flood risk.

NFIP insurance uses one of three types of Standard Flood Insurance Policies (SFIPs). SFIPs have maximum coverage limits set by law. Any federal entity that makes, guarantees, or purchases mortgages must, by law, require property owners in the SFHA to purchase flood insurance, generally through the NFIP. In moderate risk areas, community members may purchase Preferred Risk Policies (PRPs) that offer less costly insurance. The day-to-day sale, servicing, and claims processing of NFIP policies are conducted by private industry partners. Most policies are serviced by companies that are reimbursed through the Write Your Own (WYO) Program.

The premium rate for most NFIP policies is intended to reflect the true flood risk. However, Congress has directed FEMA to subsidize flood insurance for properties built before the community's first FIRM (i.e., the pre-FIRM subsidy). In addition, FEMA "grandfathers" properties at their rate from past FIRMs to updated FIRMs through a cross-subsidy. Congress also directed the development of an Affordability Study, and a forthcoming Draft Affordability Framework, to evaluate methods for making flood insurance more affordable.

Participating communities that fail to adopt FIRMs or maintain minimum floodplain standards can be put on probation or suspended from the NFIP. In communities that do not participate in the NFIP, or have been suspended, individuals cannot purchase NFIP insurance. Individuals in these communities also face challenges receiving federal disaster assistance in flood hazard areas, and have difficulties receiving federally backed mortgages.

Congress has provided appropriations to the NFIP for some of the cost of Risk MAP. Congress also authorizes the use of premium revenues for other NFIP costs, including administration, salaries, and other expenses. NFIP premiums also include other charges, such as a Federal Policy Fee, a Reserve Fund assessment, and a surcharge to help fund the NFIP. In October 2017, Congress cancelled $16 billion of NFIP debt, making it possible for the program to pay claims for Hurricanes Harvey, Irma, and Maria. The NFIP currently owes $20.525 billion to the U.S. Treasury, leaving $9.9 billion in borrowing authority from a $30.425

billion limit in law. This debt is serviced by the NFIP and interest is paid through premium revenues.

After July 31, 2018, key authorities of the NFIP, such as the authority to issue new insurance contracts, will expire if they are not reauthorized by Congress.

INTRODUCTION

The National Flood Insurance Program (NFIP) was created by the National Flood Insurance Act of 1968 (NFIA, Title XIII of P.L. 90-448, as amended, 42 U.S.C. §4001 et seq.). The NFIP received a short-term reauthorization through December 8, 2017,[1] a second short-term reauthorization through December 22, 2017,[2] and a third short-term reauthorization through January 19, 2018.[3] The NFIP lapsed between January 20 and January 22, 2018, and received a fourth short-term reauthorization until February 8, 2018.[4] The NFIP lapsed for approximately eight hours during a brief government shut-down in the early morning of February 9, 2018, and was then reauthorized until March 23, 2018.[5] The NFIP received a sixth reauthorization until July 31, 2018.[6] Prior to that, the NFIP was reauthorized by the Biggert-Waters Flood Insurance Reform Act of 2012 (Title II of P.L. 112-141, hereinafter BW-12) from July 6, 2012, to September 30, 2017. Congress amended elements of BW-12, but did not extend the NFIP's authorization further, in the Homeowner Flood Insurance Affordability Act of 2014 (P.L. 113-89, hereinafter HFIAA). The NFIP is managed by the Federal Emergency Management Agency (FEMA), through its subcomponent the Federal Insurance and Mitigation Administration (FIMA). As of February 2018, the NFIP had over 5 million flood insurance policies providing nearly $1.28 trillion in coverage. The program collects about $3.6 billion in

---

[1] P.L. 115-56, Division D, §130.
[2] P.L. 115-90.
[3] P.L. 115-96.
[4] P.L. 115-120.
[5] P.L. 115-123.
[6] P.L. 115-141, Division M, Title III.

annual premium revenue.[7] Nationally, as of March 2018, about 22,315 communities in 56 states and jurisdictions participated in the NFIP.[8] According to FEMA, the program saves the nation an estimated $1.87 billion annually in flood losses avoided because of the NFIP's building and floodplain management regulations.[9]

This chapter provides introductory information on key components of the NFIP, ranging from floodplain mapping to the standard flood insurance forms. This chapter will be updated as significant revisions are made to the NFIP through legislation or administrative action. However, this chapter does not provide detail on current or future legislative issues for Congress, which are covered in a separate report.[10] CRS also has a separate report on flood insurance and other federal disaster assistance programs.[11]

PURPOSE OF THE NFIP

In the original NFIP statute, Congress stipulated that "a program of flood insurance can promote the public interest by providing appropriate protection against the perils of flood losses and encouraging sound land use by minimizing exposure of property to flood losses."[12] Congress had found that postdisaster flood losses, and the subsequent federal disaster relief assistance to help communities recover from those flood losses, had

---

[7] Statistics on the NFIP policy and claims are available from FEMA's website at https://www.fema.gov/policy-claim-statistics-flood-insurance.

[8] Based on FEMA's map inventory, 98.8% of the U.S. population is mapped with an existing flood map. Over 88% of the population lives in a community that has received a modernized product. Email correspondence from FEMA Congressional Affairs staff, April 20, 2017. Detailed information about which communities participate and where is available from the Community Status Book, found on FEMA's website at https://www.fema.gov/national-flood-insurance-program-community-status-book.

[9] Ibid.

[10] For more information on current legislation and issues for Congress, see CRS Report R45099, *National Flood Insurance Program: Selected Issues and Legislation in the 115th Congress*, by Diane P. Horn.

[11] CRS Report R44808, *Federal Disaster Assistance: The National Flood Insurance Program and Other Federal Disaster Assistance Programs Available to Individuals and Households After a Flood*, by Diane P. Horn.

[12] See 82 Stat. 573 for text in original statute (Section 1302(c) of P.L. 90-448). This language remains in statute (see 42 U.S.C. §4001(c)).

"placed an increasing burden on the Nation's resources" and that as a matter of national policy "a reasonable method of sharing the risk of flood losses is through a program of flood insurance which can complement and encourage preventive and protective measures."[13] At the time of establishment of the NFIP, as is generally still the case today, it was found that "many factors have made it uneconomic for the private insurance industry alone to make flood insurance available to those in need of such protection on reasonable terms and conditions."[14]

Thus, the NFIP essentially has two interrelated policy purposes that can be summarized as:

1) to provide access to primary flood insurance, thereby allowing for the transfer of some of the financial risk of property owners to the federal government, and
2) to mitigate and reduce the nation's comprehensive flood risk[15] through the development and implementation of floodplain management standards.

A core design feature of the NFIP is that communities[16] are not required to participate in the program by any law or other regulation. Rather, communities in the United States voluntarily participate in the NFIP generally as a means of securing access to the primary flood insurance offered by the NFIP. Essentially, the NFIP is structured so that the availability of primary flood insurance through the NFIP (purpose #1 from above) is tied to the adoption and enforcement of floodplain management standards by participating communities (purpose #2). FEMA

---

[13] See 82 Stat. 573 for text in original statute (Section 1302(a) of P.L. 90-448). This language remains in statute (see 42 U.S.C. §4001(a)).

[14] See 82 Stat. 573 for text in original statute (Section 1302(b)(1) of P.L. 90-448). This language remains in statute (see 42 U.S.C. §4001(b)(1)).

[15] In the context of this report, *comprehensive* flood risk means that the risk includes both financial risk (i.e., physical damage to property), but also the risk to human life.

[16] 44 C.F.R. 59.1 defines community as

> any State or area or political subdivision thereof, or any Indian tribe or authorized tribal organization, or Alaska Native village or authorized native organization, which has authority to adopt and enforce flood plain management regulations for the areas within its jurisdiction.

is only allowed to provide flood insurance to "those States or areas (or subdivisions thereof)" where "adequate land use and control measures" have been adopted that "are consistent with the comprehensive criteria for land management and use developed" by the NFIP.[17] Thus, communities that participate in the NFIP, and therefore whose residents may access the NFIP's primary flood insurance, also adopt through local or state laws minimum floodplain management standards that are described in FEMA regulations.

Reduction of Comprehensive Flood Risk

The NFIP accomplishes the goal of reducing comprehensive flood risk primarily by requiring participating communities to

- Collaborate with FEMA to develop and adopt flood maps called Flood Insurance Rate Maps (FIRMs).
- Enact minimum floodplain standards based on those flood maps.

In addition, premiums collected from the sale of insurance in the NFIP finance a Flood Mitigation Assistance (FMA) grant program that reduces overall flood risk. This section of the chapter briefly discusses each of these means of reducing comprehensive flood risk.

Risk Mapping, Assessment, and Planning (Risk MAP) and Flood Insurance Rate Maps (FIRMs)

FEMA is responsible for undertaking Flood Insurance Studies (FISs) nationwide to identify areas within the United States having special flood, mudslide, and flood-related erosion hazards; assess the flood risk; and

[17] 42 U.S.C §4012(c)(2).

designate insurance zones.[18] FEMA develops, in coordination with participating communities, flood maps called Flood Insurance Rate Maps (FIRMs) using these FISs that depict the community's flood risk and floodplain. In BW-12, Congress revised the authorities of FEMA as it relates to flood hazard mapping to formally establish what FEMA has called the Risk Mapping, Assessment and Planning (Risk MAP) process.[19] Though formally authorized in BW-12, FEMA started the Risk MAP process at the request of Congress in 2009.[20] While FEMA is largely responsible for the creation of the FIRM, the community itself must pass the map into its local or state law in order for the map to be effective.

Flood Zones

An area of specific focus on the FIRM is the Special Flood Hazard Area (SFHA). The SFHA is intended to distinguish the flood risk zones that have a chance of flooding during a "1 in 100 year flood" or greater frequency. This means that properties have a risk of 1% *or greater* risk of flooding every year if in the SFHA. Table 1 shows flood-risk zones that are depicted on the FIRMs. Zones A (A1-30), AE, AH, AO, V, VE, VO, and V1-30 constitute the designated SFHA on the community's FIRM. V zones are distinguished from A zones in that V zones are subject to tidal wave action (i.e., coastal flooding). Two other designations for classifying zones in the SFHA are the Zone AR, which is an area where a levee or similar structure is determined not to provide sufficient flood protection, but is undergoing restoration; and the Zone A99, an area where a federal flood protection structure is under construction to provide the necessary flood protection standard.

---

[18] See 42 U.S.C. §4101 and 44 C.F.R. Part 65.

[19] §100216 of P.L. 112-141, 126 Stat. 927, as codified at 42 U.S.C. §4101b.

[20] Congress called for the creation of new a five year National Flood Map Maintenance Plan for fiscal years 2010-2014 in the Explanatory Statement which accompanied the Department of Homeland Security Appropriations Act, 2009 (P.L. 110-329). For the initial Risk MAP plan, see FEMA, *Risk Mapping, Assessment, and Planning (Risk MAP) Multi-Year Plan: Fiscal Years 2010-2014*, March 16, 2009, at http://www.fema.gov/media-library-data/20130726-1650-20490-4732/fema_risk_map_plan.pdf.

Table 1. Flood Zones as Depicted on Flood Insurance Rate Maps (FIRMs)

Zone Symbol	Description
A	Area of special flood hazard without water surface elevations determined.
A1-30, AE	Area of special flood hazard with water surface elevations determined.
AO	Area of special flood hazards having shallow water depths and/or unpredictable flow paths between 1 and 3 ft.
A99	Area of special flood hazard where enough progress has been made on a protective system, such as dikes, dams, and levees, to consider it complete for insurance rating purposes.
AH	Areas of special flood hazards having shallow water depths and/or unpredictable flow paths between 1 and 3 feet, and with water surface elevations determined.
AR	Area of special flood hazard that results from the decertification of a previously accredited flood protection system that is determined to be in the process of being restored to provide base flood protection.
V	Area of special flood hazards without water surface elevations determined, and with velocity, that is inundated by tidal floods (coastal high hazard area).
V1-30, VE	Area of special flood hazards, with water surface elevations determined and with velocity, that is inundated by tidal floods (coastal high hazard area).
VO	Area of special flood hazards having shallow water depths and/or unpredictable flow paths between 1 and 3 ft. and with velocity.
B, X	Areas of moderate flood hazards or areas of future-conditions flood hazard.
C, X	Area of minimal hazards.
D	Area of undetermined, but possible, flood hazards.
M	Area of special mudslide (i.e., mudflow) hazards.
N	Area of moderate mudslide (i.e., mudflow) hazards.
P	Area of undetermined, but possible, mudslide hazards.
E	Area of special flood-related erosion hazards.

Source: Adapted from 44 C.F.R. §60.3 by CRS.

Updating Flood Maps

Flood maps adopted across the country vary considerably in age and in quality. While some FIRMs may have last been developed and adopted by a community in the 1980s, especially in rural areas of the country, most communities will have maps adopted within the past 15 to 20 years.[21] All official FIRMs can be accessed, and are searchable by address and location, on a FEMA website called the Map Service Center,[22] and modern FIRMs can be digitally viewed via the Geographic Information System in the National Flood Hazard Layer.[23]

There is no consistent, definitive timetable for when a particular community will have their maps revised and updated. FEMA uses a process called the Coordinated Needs Management Strategy to prioritize, identify, and track the lifecycle of mapping needs of Risk MAP.[24] Generally, flood maps may require updating when there have been significant new building developments in or near the flood zone, changes to flood protection systems (e.g., levees and sand dunes), and environmental changes in the community. Because of the variability in how and when a FIRM is updated, for example, one community may be undergoing the process of updating its map while a neighboring community is not, and one community may have had its map last updated in 2016 while a neighboring community had its last revised in 2005, etc.

There are statutory guidelines for how FEMA is allowed to develop new FIRMs for a community. These guidelines require, for example, FEMA to conduct extensive communication and outreach efforts with the community during the mapping process and include various minimum

---

[21] By law, FEMA is required to "assess the need to revise and update all floodplain areas and flood risk zones" every five years, but not necessarily update the maps. See 42 U.S.C. §4101(e).

[22] See the Map Service Center website at https://msc.fema.gov/portal. In addition, one can review the last revision date of a community's FIRM in the Community Status Book found at https://www.fema.gov/national-flood-insurance-program-community-status-book.

[23] For more on the NFHL, and directions for accessing it, see FEMA's website at https://www.fema.gov/national-flood-hazard-layer-nfhl.

[24] For more, see FEMA, *Coordinated Needs Management Strategy Fact Sheet*, April 2015, at https://www.fema.gov/ media-library/assets/documents/21436.

waiting periods after intermediary steps are taken in the process.[25] In addition, during this process, communities are asked to submit pertinent data concerning their flood hazards, flooding experience, mitigation plans to avoid potential flood hazards, and estimates of historical and prospective economic impacts flooding has had on the community.[26] Generally, FEMA seeks to make the Risk Map a collaborative process with local communities to encourage a joint sense of "ownership" of the maps. There are also legal requirements allowing communities and individuals to appeal during the process of updating FIRMs.[27] This appeal process now includes the option, first authorized in BW-12, for communities to appeal to a Scientific Resolution Panel regarding a proposed FIRM.[28]

In BW-12, Congress reestablished and reauthorized a council called the Technical Mapping Advisory Council (TMAC).[29] The TMAC is broadly authorized to review and recommend improvements to how FEMA produces and disseminates flood hazard, flood risk, and flood map information.[30] In particular, the TMAC is authorized to recommend to FEMA "mapping standards and guidelines for—(A) flood insurance rate maps [FIRMs]; and (B) data accuracy, data quality, data currency, and data eligibility."[31] Currently, TMAC estimates that the production of a new or revised FIRM is designed to take three to five years under the Risk MAP program, but can often take as long as six and a half years or longer.[32] The

---

[25] See, for example, 42 U.S.C. §4101b(d)(1) and 42 U.S.C. §4104.

[26] See, for example, 44 C.F.R. §66.1.

[27] Primarily, see 42 U.S.C. §4104(c)-(g).

[28] §100218(a) of P.L. 112-141, 126 Stat. 930, as codified at 42 U.S.C. §4104-1. For more on the Scientific Resolution Panel, see the Panel's website at http://www.floodsrp.org/.

[29] Section 100215, Title II of P.L. 112-141, 126 Stat. 924, as codified at 42 U.S.C. §4101a. Congress originally authorized the creation of the TMAC in 1994 (see Section 576 of P.L. 103-325, 108 Stat. 2280). However, in that originating statute, the TMAC was required to terminate "5 years after the date on which all members of the Council have been appointed." BW-12 did not include a termination clause for TMAC, thus making it permanent. BW-12 describes the conditions for membership, pay, and other matters relating to the operations and structure of the TMAC.

[30] For a list of duties, see 42 U.S.C. §4101a(c).

[31] 42 U.S.C. §4101a(c)(2).

[32] For further details on the remapping process, see Section 4.4.2 in Technical Mapping Advisory Council, *Annual Report, 2015*, December 2015, pp. 4-55, and the section on Process, in Technical Mapping Advisory Council, *National Flood Mapping Program Review,* June 2016, pp. 13-17, both at http://www.fema.gov/media-library/assets/documents/ 111853.

TMAC has suggested that the ideal Risk MAP project timeline is 25 months.[33]

Map Corrections

After a map is finalized and adopted by a community, it can still be revised to correct for errors in map accuracy. To correct these inaccuracies, FEMA allows individuals and communities to request letters amending or revising the flood map. In general, there are two primary circumstances that may result in changes to the flood map. First, the natural elevation of property may be incorrectly accounted for on a FIRM, and that natural elevation is such that the property should not be considered as part of the SFHA. Generally, in this circumstance, an individual or community may request a Letter of Map Amendment (LOMA).[34] Second, a community may feel that a physical development in the community has resulted in a reduction of the flood risk for areas previously mapped in the floodplain. Generally, in this circumstance, the community may request a Letter of Map Revision (LOMR).[35] In either a LOMA or LOMR, the decision to correct a map must be based on scientific information validating the inaccuracy of the current map. In most circumstances, the cost of requesting the map correction is borne by the community or individual.[36]

State and Local Land Use Controls

As authorized in law, FEMA has developed a set of minimum floodplain management standards that are intended to

---

[33] Ibid. See Figure 4-10.

[34] For more on LOMAs, see 44 C.F.R. Part 70, or FEMA's website at http://www.fema.gov/letter-map-amendment-loma.

[35] For more on LOMRs, see 44 C.F.R. Part 72, or FEMA's website at http://www.fema.gov/letter-map-revision.

[36] For detailed information on the process of revising and amending NFIP maps, please see FEMA, *Appeals, Revisions, and Amendments to National Flood Insurance Program Maps: A Guide for Community Officials*, December 2009, at https://www.fema.gov/media-library/assets/documents/17930.

> (1) constrict the development of land which is exposed to flood damage where appropriate, (2) guide the development of proposed construction away from locations which are threatened by flood hazards, (3) assist in reducing damage caused by floods, and (4) otherwise improve the long-range land management and use of flood-prone areas.[37]

Communities are required to adopt these minimum floodplain management standards in order to participate in the NFIP.[38] FEMA has set forth the minimum standards it requires for participation in the NFIP in federal regulations.[39] Though the standards appear in federal regulations, the standards only have the force of law because they are adopted and enforced by a state or local government. Key conditions of the NFIP minimum standards include, among many other conditions, that communities

- require permits for development in the SFHA;
- require elevation of the lowest floor of all new residential buildings in the SFHA to or above the Base Flood Elevation (BFE);
- restrict development in the regulatory floodway to prevent increasing the risk of flooding; and
- require certain construction materials and methods that minimize future flood damage.[40]

---

[37] 42 U.S.C. §4102(c).

[38] 42 U.S.C. 4022(a)(1).

[39] See 44 C.F.R. Part 60, particularly 44 C.F.R. 60.3.

[40] For more on the NFIP minimum floodplain standards, see, for example:

FEMA, *NFIP Floodplain Management Requirements: A Study Guide and Desk Reference*, at https://www.fema.gov/ floodplain-management-requirements.

Christopher P. Jones, William L. Coulbourne, and Jamie Marshall, et al., Evaluation of the National Flood Insurance Program's Building Standards: Prepared as part of the 2001-2006 Evaluation of the National Flood Insurance Program, American Institutes of Research, October 2006, at https://www.fema.gov/media-library-data/20130726-1602-20490-5110/nfip_eval_building_standards.pdf.

Association of State Floodplain Managers, *A Guide for Higher Standards in Floodplain Management*, March 2013, at https://www.floods.org/ace-files/documentlibrary/ committees/3-13_Higher_Standards_in_Floodplain_Management2.pdf.

Legal enforcement of the floodplain management standards is the responsibility of the participating NFIP community. However, FEMA, often in cooperation with state governments, will conduct community assistance visits (CAVs) to monitor how and if a community is adequately enforcing its floodplain ordinances.[41] Two previous reviews commissioned by FEMA on community enforcement of minimum floodplain standards have estimated that the nationwide rate of community compliance with the standards is 70% to 85%,[42] and that between 58% and 70% of buildings are built in full compliance with the standards.[43] A community that has been found failing to enforce the floodplain management standards may be placed on probation and ultimately suspended from the NFIP (as discussed later in this chapter).[44] As these standards are just *minimum* requirements, states and communities can elect to adopt higher standards as a means of mitigating flood risk. In addition, FEMA operates a program, called the Community Rating System, to incentivize NFIP communities to adopt more rigorous floodplain management standards (as discussed later in this chapter).[45]

---

[41] For more information on CAVs, see FEMA, *Guidance for Conducting Community Assistance Contacts and Community Assistance Visits*, F-776, April 2011, at https://www.fema.gov/media-library-data/20130726-1812-25045-9789/fema_f776_cacs_cavs_web_final_apr2011.pdf.

[42] A community was estimated to be compliant with the floodplain management standards if it had no program deficiencies or violations or if it addressed them satisfactorily within two years [following a community assistance visit]. As another way of gauging overall community compliance, FEMA and state personnel were asked in interviews to give estimates of the proportion of compliant communities in their 'territories.' Their responses ranged from 0 percent compliant for some areas to 100 percent for others, but averaged to 78 percent compliant, the median of the range calculated from existing records.
See Jacquelyn L. Monday, Kristen Y. Grill, and Paul Esformes, et al., *An Evaluation of Compliance with the National Flood Insurance Program Part A: Achieving Community Compliance*, American Institutes of Research, Prepared as part of the 2001-2006 Evaluation of the National Flood Insurance Program, November 2006, p. x, at https://www.fema.gov/media-library-data/20130726-1602-20490-1461/nfip_eval_community_compliance_a.pdf.

[43] Margaret L. Mathis and Suzanne Nicholson, *An Evaluation of Compliance with the National Flood Insurance Program Part B: Are Minimum Building Requirements Being Met?*, Dewberry, Prepared as part of the 2001-2006 Evaluation of the National Flood Insurance Program, October 2006, p. viii, at https://www.fema.gov/media-library-data/20130726-1602-20490-2430/nfip_eval_community_compliance_b.pdf.

[44] See the "Nonparticipating Communities and Community Suspension" section of this report.

[45] See the "Community Rating System" section of this report.

Flood Mitigation Assistance Grants

To reduce comprehensive flood risk, FEMA also operates a Flood Mitigation Assistance (FMA) Grant Program that is funded through revenue collected by the NFIP.[46] The FMA Program[47] awards grants for a number of purposes, including state and local mitigation planning; the elevation, relocation, demolition, or flood proofing of structures; the acquisition of properties; and other activities.[48] In FY2014, the FMA Program was authorized to use $100 million of NFIP revenue. In FY2015 it was authorized to use $150 million; in FY2016, $175 million; in FY2017, $175.1 million; and in FY2018, $175 million.[49] The funding is available until it is expended, so in certain years the amount awarded may exceed the amount authorized by Congress in an appropriation act for a specific fiscal year.[50] A database of approved FMA grants that is available from FEMA indicates that over $905 million in projects was approved between July 1997 and November 2017.[51]

---

[46] 42 U.S.C. §4104c.

[47] In BW-12, Congress mandated that the grant assistance previously delivered by the Repetitive Flood Claims (RFC) and the Severe Repetitive Loss (SLR) grant programs should be unified into a single program, FMA, by rescinding the authorization for the SLR program and the RFC program. See Sections 100225(b)-(c) of P.L. 112-114, respectively.

[48] For additional information on the FMA Program, see 44 C.F.R. Part 78, FEMA's website at https://www.fema.gov/ flood-mitigation-assistance-grant-program, and FEMA, *FY 2016 Flood Mitigation Assistance (FMA) Grant Program Fact Sheet*, February 15, 2016, at http://www.fema.gov/media-library-data/1455710459301-048a67862580037b30cd640a802a9053/FY16_FMA_Fact_Sheet.pdf.

[49] See, respectively, P.L. 113-76, 128 Stat. 265; P.L. 114-4, 129 Stat. 58; P.L. 114-113, 129 Stat. 2508, P.L. 115-31, 131 Stat. 417, and P.L. 115-141.

[50] For additional information on how the FMA program is financed, see the "Appropriations and Offsetting Receipts" section of this report.

[51] This figure represents the total amount of federal assistance, without subtracting the cost share, for the three flood mitigation programs that existed during this time: SRL, RFC, and FMA. To access the database, see FEMA's website at https://www.fema.gov/media-library/assets/documents/103339.

PRIMARY FLOOD INSURANCE THROUGH THE NFIP

Standard Flood Insurance Policies (SFIPs)

FEMA has considerable discretion under the law to craft the details of the flood insurance policies it sells through the NFIP.[52] Currently, there are three policies that the NFIP uses to sell primary flood insurance—the Dwelling, the General Property, and the Residential Condominium Building Association policy forms. Collectively, these Standard Flood Insurance Policies (SFIPs) appear in regulations, and coverage qualifications are generally equivalent.[53] Table 2 displays the maximum available coverage limits for SFIPs by occupancy type. These coverage amounts are set by law.[54] Policyholders are able to elect coverage for both their building property and separate coverage for contents. Renters may obtain contents-only coverage.

Table 2. Maximum Available Coverage Limits for SFIPs by Occupancy Type

	Occupancy Type		
Policy Coverage Type	1-4 Family	Other Residential	Nonresidential Business, other Nonresidential
Combined Building/ Contents	$250,000/$100,000	$500,000/$100,000	$500,000/$500,000
Contents Only	$100,000	$100,000	$500,000

Source: FEMA, *Flood Insurance Manual, Rating Section*, revised April 2018, p. RATE 1, at https://www.fema.gov/ media-library/assets/documents/162601.

[52] 42 U.S.C. §4013(a).

[53] See 44 C.F.R. Part 61, Appendix A. Copies of the policy forms are also available on FEMA's website at https://www.fema.gov/national-flood-insurance-program/standard-flood-insurance-policy-forms.

[54] 42 U.S.C. §4013(b).

Because SFIP coverage limits are often less than the value of a structure or the value of the property's contents, policyholders can obtain excess flood insurance to cover losses beyond the coverage limit. However, such excess coverage is not sold by the NFIP, and can only be purchased through the private insurance market.

Within the SFIPs sold by the NFIP, there are numerous policy exclusions that are often not understood by policyholders. For example, SFIPs do *not* provide coverage for alternative living expenses (e.g., the cost of staying in a hotel while a house is being repaired) or business interruption expenses, and SFIPs have limited coverage of basements or crawlspaces.[55] In addition, the SFIP does not cover damage caused by earth movement, including landslides.[56]

Mandatory Mortgage Purchase Requirement

In a community that participates or has participated in the NFIP, owners of properties in the mapped SFHA[57] are required to purchase flood insurance as a condition of receiving a federally backed mortgage. By law and regulation, federal agencies, federally regulated lending institutions, and government-sponsored enterprises must require these property owners to purchase flood insurance as a condition of any mortgage that these entities make, guarantee, or purchase.[58] Examples of the types of lenders that are mandated to issue regulations requiring the purchase of flood insurance related to mortgages include

---

[55] For a basic guide to coverage under the SFIP, see FEMA, *Summary of Coverage*, FEMA F-679, November 2012, at http://www.fema.gov/media-library-data/20130726-1620-20490-4648/f_679_summaryofcoverage_11_2012.pdf.

[56] The exclusion for earth movement was particularly contentious in the aftermath of Hurricane Sandy. For a basic discussion, see Lloyd Dixon, Noreen Clancy, and Bruce Bender, et al., *Flood Insurance in New York City Following Hurricane Sandy*, RAND Corporation, October 2013, at http://www.rand.org/pubs/research_reports/RR328.html.

[57] For additional information on the SFHA, see the "Risk Mapping, Assessment, and Planning (Risk MAP) and Flood Insurance Rate Maps (FIRMs)" section of this report.

[58] 42 U.S.C. §4012a.

- federal agency lenders, such as the Department of Veterans Affairs, or
- the government-sponsored enterprises (GSEs), Federal National Mortgage Association (Fannie Mae) and Federal Home Loan Mortgage Corporation (Freddie Mac), or
- federally regulated lending institutions, such as banks covered by the Federal Deposit Insurance Corporation (FDIC) or the Office of the Comptroller of the Currency (OCC).[59]

Property owners falling under this mandate may purchase flood insurance through the NFIP, or through a private company, so long as the private flood insurance "provides flood insurance coverage which is at least as broad as the coverage" of the NFIP, among other conditions.[60]

Not all mortgages in the SFHA are affected by this mandatory purchase requirement. For example, a personal mortgage loan between two private parties (such as between family members), or a mortgage issued by a private mortgage company that is *not* then sold on the secondary market to a bank or entity like Fannie Mae, *may* not require flood insurance. Even if they are not technically required to mandate flood insurance by federal law, the issuing party may still require it as a means of financially securing the property. While the exact percentage of total mortgages requiring flood insurance is unknown, one study suggested at least 77% of all mortgages in SFHAs in 2003 would be subject to the requirement.[61]

---

[59] 42 U.S.C §4012a(b)(1)-(3). For examples of regulations and policies implementing this provision, see 12 C.F.R. Parts 12 and 339 for regulations for the OCC and FDIC, respectively, or Federal National Mortgage Association (Fannie Mae), *Servicing Guide, Part B-3: Flood Insurance Requirements*, 2015, available at https://www.fanniemae.com/content/guide/servicing/b/index.html.

[60] 42 U.S.C §4012a(b). For additional information on private flood insurance, see CRS Insight IN10450, *Private Flood Insurance and the National Flood Insurance Program (NFIP)*, by Baird Webel and Diane P. Horn.

[61] The 77% figure was estimated using the overall percentage of mortgages held by parties subject to the mortgage requirement, and likely is subject to a degree of error. See Richard J. Tobin and Corinne Calfee, *The National Flood Insurance Program's Mandatory Purchase Requirement: Policies, Process, and Stakeholders*, American Institutes for Research, Prepared as part of the Evaluation of the National Flood Insurance Program, March 2005, p. 31, available at http://www.fema.gov/media-library-data/20130726-1602-20490-9257/ nfip_eval_mandatory_purchase_requirement.pdf.

Despite the mandatory purchase requirement, not all covered mortgages carry the insurance as dictated. Though there are no official statistics available from the federal mortgage regulators responsible for implementation of the mandate, a 2006 study commissioned by FEMA found that compliance with this mandatory purchase requirement may be as low as 43% in some areas of the country (the Midwest), and as high as 88% in others (the West).[62] In a 2013 analysis done following Hurricane Sandy, one study found that approximately 65% of properties in New York City required to have insurance through their mortgage had such insurance.[63] A 2017 study of flood insurance in New York City by the same authors reassessed the 2013 data and suggested that the estimate in their earlier study may have slightly overstated the actual take-up rate, which the 2017 study estimated at 61%. The later study found that compliance with the mandatory purchase requirement by properties in the SFHA with mortgages increased from 61% in 2012 to 73% in 2016. The later study also argued that findings for properties without mortgages indicate the effectiveness of the mandatory purchase requirement, as the 37% take-up rate for properties without mortgages in the SFHA was similar to take-up rates outside the SFHA (37% for properties with mortgages and 32% for properties without mortgages).[64]

The escrowing of insurance premiums may increase compliance with the mandatory purchase requirement. Federal mortgage regulators have required the escrowing of flood insurance premiums on certain mortgages in compliance with regulations issued after changes to the law made in 1994.[65] Expanding upon existing requirements, Section 100209 of BW-

---

[62] Lloyd Dixon, Noreen Clancy, and Seth A. Seabury, et al., *The National Flood Insurance Program's Market Penetration Rate: Estimates and Policy Implications*, RAND Corporation, Prepared as part of the Evaluation of the National Flood Insurance Program, February 2006, p. 23, available at https://www.fema.gov/media-library-data/ 20130726-1602-20490-2804/nfip_eval_market_penetration_rate.pdf.

[63] Lloyd Dixon, Noreen Clancy, Bruce Bender, et al., *Flood Insurance in New York City Following Hurricane Sandy*, RAND Corporation, 2013, p. 15, available at http://www.rand.org/pubs/research_reports/RR328.html.

[64] Lloyd Dixon, Noreen Clancy, and Benjamin M. Miller, et al., The Cost and Affordability of Flood Insurance in New York City: Economic Impacts of Rising Premiums and Policy Options for One- to Four-Family Homes, RAND Corporation, March 2017, pp. 15-18, https://www.rand.org/pubs/research_reports/RR1776.html.

[65] Section 523 of P.L. 103-325, 108 Stat. 2258.

12,[66] as subsequently revised by Section 25 of HFIAA,[67] has required that regulated lenders start escrowing flood insurance for all mortgages, except if the lending institution is under a regulated size or the loan is a subordinate to another loan. This broader implementation of the escrowing provision began in January 2016, per law and regulations.[68]

Preferred Risk Policies (PRPs)

Flood insurance is optional for properties outside the SFHA regardless of whether they have a federally backed mortgage. However, as there is still a risk of flooding outside the SFHA, members of NFIP participating communities with property located in the B, C, or X Zones of a FIRM may voluntarily purchase a lower-cost Preferred Risk Policy. Unlike with properties in the SFHA, an individual may be denied a PRP if there is significant loss history for the property.[69] FEMA encourages the purchase of PRPs both to reduce the financial flood risk of a broader group of individuals, and to expand the policy base of the NFIP writ large, thus improving the fiscal soundness of the NFIP portfolio. A PRP uses the same basic policy forms as properties within the SFHA, but receives discounted rates in accordance with its lower risk profile.

---

[66] Section 100209 of P.L. 112-141, 126 Stat. 920.

[67] Section 25 of P.L. 113-89, 128 Stat. 1030.

[68] See 42 U.S.C 4012a(d)(1), and the final rule of Office of the Comptroller of the Currency, Treasury; Board of Governors of the Federal Reserve System; Federal Deposit Insurance Corporation; Farm Credit Administration; National Credit Union Administration, "Loans in Areas Having Special Flood Hazards," 80 *Federal Register* 43216, July 21, 2015. For a full explanation of how escrowing had previously worked prior to BW-12, and for how it is being required following BW-12 and HFIAA, see the notice of proposed rulemaking from the federal agency regulators at Office of the Comptroller of the Currency, Treasury; Board of Governors of the Federal Reserve System; Federal Deposit Insurance Corporation; Farm Credit Administration; National Credit Union Administration, "Loans in Areas Having Special Flood Hazards," 79 *Federal Register* 64518, October 29, 2014.

[69] For additional information on PRPs and restrictions, see FEMA, *Flood Insurance Manual, Preferred Risk Policy Section*, revised April 2018, at https://www.fema.gov/media-library-data/1523307331106- 4cf9726b2eb04c3471a3e9d37a58fa6a/09_prp_508_apr2018.pdf.

Increased Cost of Compliance (ICC) Coverage

The NFIP requires most SFIP and PRP policyholders[70] to purchase what is in effect a separate insurance policy to offset the expense of complying with more rigorous building code standards when local ordinances require them to do so. This increased cost of compliance coverage is authorized in law, and rates for the coverage, as well as how much can be paid out for claims, are set by FEMA.[71] Congress has capped the amount that can be paid for ICC coverage at $75.[72] The ICC policy has a separate rate premium structure, and provides an amount up to $30,000 in payments for certain eligible expenses.[73]

For example, when a building is determined by a community to be substantially damaged[74] following a flood, floodplain management standards adopted by local communities can require the building to be rebuilt to current floodplain management requirements, even if the property previously did not need to do so. For instance, the new compliance standard may require the demolition and elevation of the rebuilt building to above the BFE. An ICC claim may then be submitted by the policyholder to offset the cost of complying with the elevation standard. FEMA also makes ICC coverage available if a building has been declared a repetitive loss[75] by a community's floodplain management regulations.[76] However, not all participating NFIP communities have or

---

[70] For example, ICC coverage is not required on condominium units and content-only policies.

[71] 42 U.S.C. §4011(b).

[72] Ibid.

[73] For example, for ICC premiums, see FEMA, *Flood Insurance Manual, Rating Section*, Revised April 2018, p. RATE 19, at https://www.fema.gov/media-library-data/1523307287100-4cf9726b2eb04c3471a3e9d37a58fa6a/ 05_rating_508_apr2018.pdf.

[74] 44 C.F.R §59.1 defines "substantial damage" as damage of any origin sustained by a structure whereby the cost of restoring the structure to its before damaged condition would equal or exceed 50% of the market value of the structure before the damage occurred.

[75] 42 U.S.C. §4121(a)(7) defines repetitive loss structure as

a structure covered by a contract for flood insurance that—

(A) has incurred flood-related damage on 2 occasions, in which the cost of repair, on the average, equaled or exceeded 25 percent of the value of the structure at the time of each such flood event; and

(B) at the time of the second incidence of flood-related damage, the contract for flood insurance contains increased cost of compliance coverage.

[76] 42 U.S.C §4011(b)(1).

enforce a "repetitive loss provision" that records, declares, and mandates improvements to properties that have experienced repetitive loss. Thus, certain structures that have experienced repetitive loss may not be eligible for ICC payments.[77]

FEMA has not implemented ICC coverage for two conditions that they are authorized to do so by law. These two conditions are for properties that have sustained flood damage on multiple occasions, if the Administrator determines that it is cost-effective and in the best interests of the NFIP, and for properties for which an offer of mitigation assistance is made under various federal assistance programs.[78] FEMA's decision not to implement these provisions has provoked criticism from some stakeholders of the NFIP.[79]

Servicing of Policies and Claims Management

While FEMA provides the overarching management and oversight of the NFIP, the bulk of the day-to-day operation of the NFIP, including the marketing, sale, writing, and claims management of policies, is handled by private companies. This arrangement between the NFIP and private industry is authorized by statute and guided by regulation.[80] There are two different arrangements that FEMA has established with private industry. The first is the Direct Servicing Agent, or DSA, which operates as a private contractor on behalf of FEMA for individuals seeking to purchase flood insurance policies directly from the NFIP.[81] The second arrangement

---

[77] For additional information on repetitive loss and repetitive loss provisions, see FEMA, *Increased Cost of Compliance (ICC) Coverage: Guidance for State and Local Officials*, September 1, 2003, at http://www.fema.gov/media-library/ assets/documents/1973.

[78] See 42 U.S.C §4011(b)(3) and (4).

[79] See, for example, Association of State Floodplain Managers (ASFPM), *A Vision for Implementing ICC as it Exists in Law Today*, February 29, 2016, at http://www.floods.org/ace-images/ICC_Feb2016.pdf.

[80] See primarily 42 U.S.C. §4081 and §4018, and 44 C.F.R. Part 62.

[81] The current Direct Servicing Agent is a company called Torrent Technologies, Inc., who were awarded the contract in September 2016. See https://www.fbo.gov/index?s=opportunity&mode=form&id=58bec8ac15f6944abe778f4ae49a0841&tab=core&_cview=1. The website for Torrent Technologies, Inc. is at http://torrentcorp.com/.

is called the Write-Your-Own (WYO) Program, where private insurance companies are paid to directly write and service the policies themselves. With either the DSA or WYO Program, the NFIP retains the actual financial risk of paying claims for the policy (i.e., underwrites the policy), and the policy terms and premiums are the same.

Currently, approximately 12% of the total NFIP policy portfolio is managed through the DSA, and 88% of NFIP policies are sold by the 67 companies participating in the WYO Program.[82] Over the years, the balance between the number of policies serviced by the WYO Program or the DSA has evolved, with the WYOs covering approximately 50% of policies in 1986, and approximately 97% of policies in 2008.[83] Because most purchasers of the NFIP policies never interface directly with a FEMA representative, and only deal with a WYO company or the DSA, they may not be aware that they are actually purchasing insurance from FEMA.

Companies participating in the WYO Program are compensated through a variety of methods, as summarized in Table 3. Especially following Hurricane Katrina and Hurricane Sandy, the Government Accountability Office (GAO) and Department of Homeland Security, Office of the Inspector General (DHS IG) have investigated how much the WYOs were compensated for the services they provided in support of the NFIP.[84] In BW-12, Congress required FEMA to develop and issue a rulemaking on a "methodology for determining the appropriate amounts that property and casualty insurance companies participating in the Write Your Own program should be reimbursed for selling, writing, and

---

[82] Email correspondence from FEMA Congressional Affairs staff, February 3, 2017. A list of companies participating in the WYO Program is available at https://www.fema.gov/wyo_company.

[83] For additional background on the estimates and history of the WYO Program, see, respectively, U.S. Government Accountability Office, *Private Companies' Participation in the Write Your Own Program*, RCED-87-108, May 29, 1987, at http://www.gao.gov/products/RCED-87-108, and U.S. Government Accountability Office, *Opportunities Exist to Improve Oversight of the WYO Program*, GAO-09-455, August 2009, at http://www.gao.gov/products/GAO-09-455.

[84] For example, see U.S. Government Accountability Office, *Opportunities Exist to Improve Oversight of the WYO Program*, GAO-09-455, August 2009, at http://www.gao.gov/products/GAO-09-455 and DHS Office of Inspector General, *FEMA Does Not Provide Adequate Oversight of Its National Flood Insurance Write Your Own Program*, OIG-16-47, March 8, 2016, at https://www.oig.dhs.gov/assets/Mgmt/2016/OIG-16-47-Mar16.pdf.

servicing flood insurance policies and adjusting flood insurance claims on behalf of the National Flood Insurance Program."[85] This rulemaking was required within a year of enactment of BW-12. As of April 2018, FEMA has yet to publish a rulemaking to revise the compensation structure of the WYOs.

Table 3. Types of Compensation for WYO Companies

Compensation Type	Description
Operating and Administrative Expenses	Reimbursement of marketing, operating, and administrative expenses.
Commission Allowance	Allowance to meet commissions or salaries of insurance agents.
Growth Bonus	Increase in compensation dependent on the extent to which the Company meets the NFIP marketing goals.
Allocated Loss Adjustment Expenses (ALAE)	Standard amount of compensation for each individual claim.
Unallocated Loss Adjustment Expenses (ULAE)	General overhead (indirect expenses) for settling losses.
Special Allocated Loss Adjustment Expenses (SALAE)	Direct expenses that are above what is covered by ALAE.

Source: Table 1 of DHS Office of Inspector General, *FEMA Does Not Provide Adequate Oversight of Its National Flood Insurance Write Your Own Program*, OIG-16-47, March 8, 2016, p. 3, at https://www.oig.dhs.gov/assets/Mgmt/2016/OIG-16-47-Mar16.pdf. This replicated table is based on the OIG's analysis of the subsidy arrangement.

Following Hurricane Sandy, there were concerns raised regarding the possible systematic underpayment of claims for flood losses through the NFIP.[86] As a result of these issues, FEMA carried out a process by which Hurricane Sandy survivors could resubmit their NFIP claims to be reevaluated by FEMA. FEMA reviewed the resubmitted claims and

---

[85] Section 100224 of P.L. 112-141, 126 Stat. 936.

[86] For an analysis of the claims issues following Hurricane Sandy, see U.S. Congress, Senate Committee on Banking, Housing, and Urban Affairs, *Assessing and Improving Flood Insurance Management and Accountability in the Wake of Superstorm Sandy*, prepared by Majority Staff Report, 114th Cong., 1st sess., June 22, 2015, at https://hazdoc.colorado.edu/bitstream/handle/10590/4511/C023801.pdf?sequence=1.

provided additional claim payments to those deemed warranted in the review, and concluded the Sandy Claims Review Process on March 1, 2018.[87] As of January 29, 2018, approximately 85% of policyholders who requested a review had received additional payments, resulting in approximately $258.6 million in additional claims payments. The remaining 15% of reviewed files received no additional payment.[88] In addition, FEMA settled and litigated lawsuits initiated by claimants following Hurricane Sandy; with 1,631 of the 1,633 court cases settled, resulting in approximately $164 million in settlement payments.[89]

Pricing and Premium Rate Structure

Except for certain subsidies, flood insurance rates in the NFIP are directed to be "based on consideration of the risk involved and accepted actuarial principles,"[90] meaning that the rate is reflective to the true flood risk to the property. Essentially, FEMA uses several basic characteristics to classify properties based on flood risks. Structures are evaluated by their specific risk zone on a FIRM, the elevation of the structure relative to the Base Flood Elevation (BFE) in each risk zone,[91] and occupancy type (e.g., single family, other residential, nonresidential, and mobile/manufactured homes), along with other specific determinants of risk. In addition, the premium structure includes estimates for the expenses of the NFIP, including servicing of policies. A detailed discussion of the premium rate structure of the NFIP, and how or why it is and is not actuarially sound, is

---

[87] For figures regarding the Sandy claims review and legal settlements, see the Fact Sheet reports published by FEMA's NFIP Transformation Task Force published at https://www.fema.gov/media-library/assets/documents/107695. Since August 2016 these fact sheets have been called the Sandy Claims Review Decision Fact Sheet.

[88] Sandy Claims Review Division Update, February 1, 2018, at https://www.fema.gov/media-library-data/ 1518101164216-73461746bb31f62a38ff95a90962093f/Updated_Feb1_Final_Fact_Sheet_Sandy_Claims_Review_Division.pdf.

[89] Ibid.

[90] 42 U.S.C. §4014(a)(1).

[91] The Base Flood Elevation (BFE) is the water-surface elevation of the base flood, which is the 1%-annual-chance flood, commonly called the 100-year flood. The probability is 1% that rising water will reach BFE height in any given year. The depth of the base flood is calculated by subtracting the ground elevation from the BFE.

beyond the scope of this chapter. However, additional resources exist to assist Congress with this issue.[92]

Pre-FIRM Subsidy

While most premium rates in the NFIP are intended to represent the full flood risk of a given structure, Congress has directed FEMA *not* to charge actuarial rates for properties that were constructed or substantially improved before December 31, 1974, or before the date upon which FEMA has published the first Flood Insurance Rate Map for the community, whichever was later.[93] Therefore, by statute, premium rates charged on structures built before they were first mapped into a flood zone that have not been substantially improved, known as pre-FIRM structures, are allowed to have lower premiums than what would be expected to cover predicted claims. The availability of this pre-FIRM subsidy was intended to allow preexisting floodplain properties to contribute in some measure to prefunding their recovery from a flood disaster instead of relying solely on federal disaster assistance. In essence, the flood insurance could distribute some of the financial burden among those protected by flood insurance and the public.

As of September 2016, 817,344 policies received a pre-FIRM subsidy, representing approximately 16.1% of all NFIP policies.[94] Historically, the total number of pre-FIRM policies is relatively stable, but the percentage

---

[92] See, for example:

Several reports from GAO, including U.S. Government Accountability Office, *Continued Progress Needed to Fully Address Prior Recommendations on Rate-Setting Methods*, GAO-16-59, March 2016, at http://www.gao.gov/assets/ 680/675855.pdf.

Carolyn Kousky and Leonard Shabman, *Pricing Flood Insurance: How and Why the NFIP Differs from a Private Insurance Company*, Resources for the Future, RFF DP 14-37, October 2014, at http://www.rff.org/files/sharepoint/WorkImages/Download/RFF-DP-14-37.pdf.

National Academies of Sciences, *Affordability of National Flood Insurance Program Premiums: Part 1*, 2015, Chapter 3, National Flood Insurance Pricing, Policies, and Premiums, at http://www.nap.edu/catalog/21709/affordability-of-national-flood-insurance-program-premiums-report-1.

Congressional Budget Office, *The National Flood Insurance Program: Factors Affecting Actuarial Soundness*, November 2009, at https://www.cbo.gov/sites/default/files/cbofiles/ftpdocs/106xx/doc10620/11-04-floodinsurance.pdf.

[93] 42 U.S.C. §4015(c).

[94] Email correspondence from FEMA Congressional Affairs staff, March 3, 2017.

of those policies by comparison to the total policy base has decreased.[95] The pricing subsidy for pre-FIRM policies is progressively being phased out of the NFIP, as was initially required under Section 100205 of BW-12, as revised by Sections 3 and 5 of HFIAA.[96] Under current law, all premiums for pre-FIRM properties will eventually reach actuarially sound rates (i.e., the rate equivalent structures pay without the subsidy, reflecting true flood risk), but at a different pace of phaseout depending on the property type. Table 4 provides an adaptation of a table from GAO regarding the multifaceted phaseout of the pre-FIRM subsidy following BW-12, as revised by HFIAA. In summary, HFIAA slowed the rate of phaseout of the pre-FIRM subsidy for most primary residences, but retained the pace of the phaseout of the subsidy from BW-12 for business properties and secondary homes. In addition, HFIAA created a minimum and maximum increase in the amount for the phaseout of pre-FIRM subsidies for all primary residences of 5%-15% annually. Unless otherwise noted, the percentage increases are based on the current premium (e.g., a 15% annual increase from the prior year premium), rather than the percentage difference between the current premium and the actuarial rate (i.e., a rate increase of 25% does not mean the pre-FIRM subsidy is eliminated in four years).

Newly Mapped Subsidy

Congress introduced a new form of subsidy in HFIAA, for owners of properties newly mapped into a SFHA.[97] The newly mapped procedure applies to properties previously in zones B, C, X, D, AR, or A99,[98] which are newly mapped into a SFHA on or after April 1, 2015, if the applicant obtains coverage that is effective within 12 months of the map revision date. The newly mapped procedure does not apply to properties mapped into a SFHA by the initial FIRM for a community entering the NFIP, and

---

[95] For an historical prospective on the percentages of subsidized policies in the NFIP, see Figure 1 of U.S. Government Accountability Office, *Flood Insurance: More Information Needed on Subsidized Properties*, GAO-13-607, July 2013, p. 7, at http://www.gao.gov/assets/660/655734.pdf.

[96] P.L. 112-141, 126 Stat. 917; and P.L. 113-89, 128 Stat. 1021-1022; respectively.

[97] Section 6 of P.L. 113-89, 128 Stat.1028, as codified at 42 U.S.C. §4015(i).

[98] See Table 1.

certain properties may be excluded based on their loss history.[99] The rate for eligible newly mapped properties is equal to the PRP rate, but with a higher Federal Policy Fee,[100] for the first 12 months following the map revision. After the first year, the newly mapped rate will begin its transition to a full-risk rate, with annual increases to newly mapped policy premiums calculated using a multiplier that varies by the year of the map change.[101] Annual increases are restricted to no more than 18% per year. As of September 2016, 3.9% of all policies received a newly mapped subsidy.[102]

Grandfathering Cross-Subsidy

Using the authority to set rate classes for the NFIP and to offer lower than actuarial premiums,[103] FEMA allows property owners to maintain their old flood insurance rate class if their property is remapped into a new flood rate class. This practice is colloquially referred to as "grandfathering," "administrative grandfathering," or the "grandfather rule" and is separate and distinct from the pre-FIRM subsidy.[104] To understand the grandfather rule, consider a hypothetical property X that is currently mapped into one flood zone (e.g., Zone AE), and is built to the proper building code and standards. If property X then is remapped to a new flood zone (e.g., Zone VE) and has maintained continuous insurance coverage under the NFIP, the owner of property X can pay the flood

---

[99] For properties which are excluded from, or ineligible for, the newly mapped subsidy, see FEMA, *Flood Insurance Manual, Newly Mapped Section,* revised April 2018, pp. NM 1 and NM 2, https://www.fema.gov/media-library-data/1523307347793-4cf9726b2eb04c3471a3e9d37a58fa6a/10_newly_mapped_508_apr2018.pdf.

[100] The FRP for a newly mapped property is $50, where the FPF for PRP is $25. See FEMA, *Flood Insurance Manual, Rating Section*, Revised April 2018, p. RATE 16, at https://www.fema.gov/media-library-data/1523307287100-4cf9726b2eb04c3471a3e9d37a58fa6a/05_rating_508_apr2018.pdf.

[101] FEMA, Attachment A: Summary of the NFIP Program Changes Effective April 1, 2018, at https://nfip-iservice.com/Stakeholder/pdf/bulletin/ATTACHMENT%20A%20-%20Summary%20of%20the%20NFIP%20April%202018%20and%20January%202019%20Program%20Changes%20f inal.pdf.

[102] Email correspondence from FEMA Congressional Affairs staff, March 3, 2017.

[103] 42 U.S.C. §4013(a).

[104] For a full description, see FEMA, *NFIP Grandfathering Rules for Agents*, March 2016, at https://www.fema.gov/media-library-data/1488482596393-dcc52e6c120c9327dcd75f1c08e802e4/GrandfatheringForAgents_03_2016.pdf.

Table 4. Phaseout of NFIP Pre-FIRM Premium Subsidy Following Legislation

	Annual Rate of Phaseout	
Property Type or Circumstance	*After BW-12*	*After HFIAA (current)*
Primary residence...		
And all other properties not included in other categories	Five-year phaseout starting after property is identified on new FIRM[a]	5-15%[b]
Purchased after July 6, 2012	Eliminated immediately	Reinstated, 5-15%[b]
Not insured by NFIP as of July 6, 2012	Eliminated immediately	Reinstated, 5-15%[b]
Nonprimary residence	25%	25%
Nonresidential properties other than business properties[c]	25%	5-15%[b]
Business properties	25%	25%
Property with severe repetitive loss[d]	25%	25%
Property with substantial cumulative damage[e]	25%	25%
Property with substantial damage or improvement after July 6, 2012[f]	25%	25%
Owner let NFIP policy laps[e]	Eliminated immediately[g]	Eliminated immediately[h]
Prospective insured who refuses to accept offers for mitigation assistance[i]	Eliminated immediately	Eliminated immediately

Source: CRS analysis of P.L. 112-141 and P.L. 113-89. Adapted from Figure 3 from U.S. Government Accountability Office, *Forgone Premiums Cannot Be Measured and FEMA Should Validate and Monitor Data System Changes*, GAO-15-111, December 2014, p. 10, at http://www.gao.gov/assets/670/667413.pdf.

Notes:

[a] All properties with subsidies not being phased out at higher rates or already eliminated were required to begin paying actuarial rates following a five-year period, phased in at 20% a year, after a revised or updated flood insurance rate map (FIRM) was issued for the area containing the property. Thus, this provision would have gradually eliminated both pre-FIRM and grandfathering subsidies for all properties before being struck by HFIAA. See Section 100207 of P.L. 112-141, 126 Stat. 919.

[b] HFIAA permits individual property increases of up to 18%, but limits the rate class increases to 15% per year. See 42 U.S.C. §4015(e).

[c] Prior to BW-12, FEMA did not distinguish business properties from other nonresidential properties such as houses of worship, nonprofits, and schools. In order to implement the requirement that business properties should have their rates phased out at 25% annually, FEMA needed to increase all nonresidential properties at that rate. However, with greater distinguishing between property types possible with new data, and phaseout rate increases being capped by HFIAA, nonresidential properties other than business properties are now having their subsidies phased out at 5%-15% instead of 25%. For more discussion on this issue, see U.S. Government Accountability Office, *Status of FEMA's Implementation of the Biggert-Waters Act, as Amended*, GAO-15-178, February 2015, p. 22, at http://www.gao.gov/products/GAO-15-178, and FEMA, *Summary of the NFIP Program Changes Effective April 1, 2017*, WYO Program Bulletin W-15046, Attachment A, at http://nfipiservice.com/Stakeholder/pdf/bulletin/w-16071.html.

[d] Severe repetitive loss properties are those that have incurred four or more claim payments exceeding $5,000 each, with a cumulative amount of such payments over $20,000; or at least two claims with a cumulative total exceeding the value of the property. See 42 U.S.C. §4014(h) and 44 C.F.R. §79.2(h).

[e] Any property that has incurred flood-related damage in which the cumulative amounts of payments under the NFIP equaled or exceeded the fair market value of such property. See 42 U.S.C. §4014(a)(2)(C).

[f] Any property that has experienced or sustained substantial damage exceeding 50% of the fair market value or substantial improvement exceeding 50% of the fair market value. See 42 U.S.C. §4014(a)(2)(E). Section 100205(a)(1)(A) of BW-12 (P.L. 112-141, 126 Stat. 917) originally set the substantial improvement amount at 30%, but HFIAA raised the percentage to 50% of fair market value.

[g] As required by Section 100205(a)(1)(B) of BW-12 (P.L. 112-141, 126 Stat. 917), only for NFIP policies that lapsed in coverage as a result of the deliberate choice of the policyholder. If a policy lapses for other reasons, rate increases of 25% may have applied if the policy qualifies under one of the other property categories.

[h] HFIAA revised the condition for policy lapse to be for NFIP all policies that lapsed in coverage *unless* the decision of the policyholder to permit a lapse in coverage was as a result of the property covered by the policy no longer being required to retain such coverage. For any policy that lapses but is not automatically charged full-risk rates, rate increases of 25% or 5% to 15% may apply if the policy qualifies under one of the other categories stated above. See 42 U.S.C. §4014(g)(1).

[i] Any prospective insured who refuses to accept any offer for mitigation assistance by the FEMA (including an offer to relocate), including an offer of mitigation assistance following a Stafford Act major disaster; or in connection with a repetitive loss property or a severe repetitive loss property. See 42 U.S.C. §4014(g)(2).

insurance rate and premium based on the prior mapped zone (i.e., pay the AE rate instead of the higher VE rate). A policyholder with a property may also be grandfathered if the elevation of a base flood is changed in a map, but the property itself does not change flood zones.[106]

Congress eliminated the practice of offering grandfathering to policyholders after new maps were issued in BW-12, but then subsequently reinstated the practice in HFIAA.[107] FEMA does not have a definitive estimate on the number of properties that have a grandfathered rate in the NFIP, though data are being collected to fulfill a separate mandate of HFIAA.[108] Unofficial estimates suggest that at least 10%-20% of properties are grandfathered, and these figures may increase with time as newer maps are introduced in high population areas.[109]

FEMA does not consider the practice of grandfathering to be a subsidy for the NFIP, per se, because the discount provided to an individual policyholder is cross-subsidized by other policyholders in the NFIP. Thus, while grandfathering does intentionally allow grandfathered policyholders to pay premiums that are less than their known actuarial rate, the discount is offset by others in the same rate class as the grandfathered policyholder. Although FEMA does not have an estimate of how many properties are paying grandfathered rates, the program tries to recoup lost revenue by charging higher rates for other policies in the SFHA. It is not clear, however, whether the NFIP is increasing other SFHA policy premiums by

---

[106] For a description of the types of grandfathering, see National Academies of Sciences, *Affordability of National Flood Insurance Program Premiums: Part 1*, 2015, p. 43, at http://www.nap.edu/catalog/21709/affordability-of-national-flood-insurance-program-premiums-report-1.

[107] Section 100207 of P.L. 112-141 amended the law to require that when a property has a revised or updated flood rate class with a new flood map, the "risk premium rate charged for flood insurance on such property adjusted to accurately reflect the current risk of flood to such property" (126 Stat. 919), thus eliminating the ability to grandfather. This provision was struck by Section 4 of P.L. 113-89, 128 Stat. 1022.

[108] Section 28 of HFIAA (P.L. 113-89, 128 Stat. 1033) requires that the Administrator "clearly communicate full flood risk determinations to individual property owners regardless of whether their premium rates are full actuarial rates." To fulfill this mandate, FEMA must identify all properties that are grandfathered or pre-FIRM and notify those policyholders what their property's true flood risk is versus the risk they are currently paying for with a subsidy/cross-subsidy.

[109] Telephone correspondence with FEMA staff, January 20, 2016; and see National Academies of Sciences, *Affordability of National Flood Insurance Program Premiums: Part 1*, 2015, p. 74, at http://www.nap.edu/catalog/21709/affordability-of-national-flood-insurance-program-premiums-report-1.

an amount equal to the discount from other NFIP risk-based rates that are being paid by the grandfathered properties.[110]

Community Rating System

Through a program called the Community Rating System (CRS), FEMA encourages communities to improve upon the minimum floodplain management standards that are required to participate in the NFIP. The CRS Program, as authorized by law, is intended to incentivize the reduction of flood and erosion risk, as well as the adoption of more effective measures to protect natural and beneficial floodplain functions.[111] FEMA awards points that increase a community's "class" rating in the CRS on a scale of 1 to 10, with 1 being the highest ranking. Points are awarded for an array of improvements for how the community informs its public on flood risk; maps and regulates its floodplain; reduces possible flood damage; and provides immediate warnings and responds to flooding incidents.[112] Starting at Class 9, policyholders in the SFHA within a CRS community receive a 5% discount on their SFIP premiums, with increasing discounts of 5% per class until reaching Class 1, and at that level, policyholders in the SFHA can receive a 45% discount on their SFIP premiums. These discounts are not extended to PRPs.

In order to participate in the CRS Program, a community must apply to FEMA and document its creditable improvements through site visits and assessments. As of June 2017, FEMA estimated that only 5% of eligible NFIP communities participate in the CRS program. However, these communities have a large number of flood policies, so more than 69% of all flood policies are written in CRS-participating NFIP communities.[113]

---

[110] National Academies of Sciences, *Affordability of National Flood Insurance Program Premiums: Part 1*, 2015, p. 43, at http://www.nap.edu/catalog/21709/affordability-of-national-flood-insurance-program-premiums-report-1.

[111] 42 U.S.C. §4022(b)(1).

[112] For a full listing of possible creditable activities in the CRS, see FEMA, *NFIP Community Rating Coordinator's Manual 2017*, at https://www.fema.gov/media-library-data/1493905477815-d794671adeed5beab6a6304d8ba0b207/633300_2017_CRS_Coordinators_Manual_508.pdf.

[113] See FEMA, *Community Rating System Fact Sheet*, June 2017, at https://www.fema.gov/media-library-data/1507029324530-082938e6607d4d9eba4004890dbad39c/NFIP_CRS_Fact_Sheet_2017_508OK.pdf.

One can determine if and how highly rated a community is in the CRS Program through the most recent Flood Insurance Manual.[114]

For April 2014 premium rates, the National Research Council estimated that the CRS program provided an average 11.4% discount on SFIP premiums across the NFIP.[115] The CRS discount is cross-subsidized into the NFIP program, such that the discount for one community ends up being offset by increased premium rates in all communities across the NFIP. Therefore, for April 2014 rates, the average 11.4% discount for CRS communities was cross-subsidized and shared across NFIP communities through a cost (or load) increase of 13.4% to overall premiums.[116] Thus in some circumstances, the discount provided to communities participating in the CRS Program may be less than the expense of the overall CRS Program.

Affordability Study and Framework

Congress has expressed concern related to the perceived affordability of flood insurance premiums. In BW-12, Congress required FEMA to commission a study with the National Academy of Sciences (NAS) regarding participation in the NFIP and the affordability of premiums. The Affordability Study was not finished by its original deadline (270 days following enactment of BW-12). Congress amended the authorization for the Study while also extending the deadline in HFIAA.[117] The NAS has since completed the Affordability Study report in two parts.[118] In HFIAA,

---

[114] For example, see FEMA, *Flood Insurance Manual, Community Rating System Section*, Revised April 2018, at https://www.fema.gov/media-library-data/1523648898907-09056f549d51efc72fe60bf4999e904a/ 20_crs_508_apr2018.pdf.

[115] For additional details, see National Research Council of the National Academies, *Affordability of National Flood Insurance Program Premiums: Report 1*, 2015, pp. 44 and 114, at http://www.nap.edu/catalog/21709/affordability-of-national-flood-insurance-program-premiums-report-1.

[116] A more recent average CRS premium discount is not available; however, according to FEMA it changes very little from year to year. Email correspondence from FEMA Congressional Affairs staff, June 15, 2017.

[117] See Section 100236 of P.L. 112-141, 126 Stat. 957; as amended by Section 16 of P.L. 113-89, 128 Stat. 1026.

[118] See National Research Council of the National Academies, *Affordability of National Flood Insurance Program Premiums: Report 1*, 2015, at http://www.nap.edu/catalog/21709/ affordability-of-national-flood-insurance-program-premiums-report-1; and National Research Council of the National Academies, *Affordability of National Flood Insurance*

Congress also required FEMA to develop a Draft Affordability Framework "that proposes to address, via programmatic and regulatory changes, the issues of affordability of flood insurance sold under the National Flood Insurance Program, including issues identified in the affordability study."[119] Due 18 months following the submission of the Affordability Study, the deadline for the Framework, based on FEMA's stated date of submittal of the Affordability Study, was September 10, 2017.[120] FEMA published their Affordability Framework on April 17, 2018.[121]

NONPARTICIPATING COMMUNITIES AND COMMUNITY SUSPENSION

There are two regulatory conditions that FEMA enforces for removing a participating community from the NFIP, called probation and suspension. Whether or not a particular community has either been placed on probation or suspended can be found using the NFIP's Community Status Book.[122] Notably, a community cannot be removed from the NFIP because of increased or excess flood insurance claims and losses. Rather, probation and suspension only occur if the community fails to uphold its obligations related to floodplain management.

A community can be placed on probation by FEMA if it is found that it is failing to adequately enforce the floodplain management standards it has adopted. As established by regulations, probation can result in a fee of $50 being charged to all policyholders in the community while the community is given time to rectify FEMA's concerns regarding their implementation

---

Program Premiums: Report 2, 2016, at http://www.nap.edu/catalog/21848/affordability-of-national-flood-insurance-program-premiums-report-2.

119 Section 9(a) of P.L. 113-89, 128 Stat. 1024.

120 Section 9(c) of P.L. 113-89, 128 Stat. 1024. FEMA has stated it officially submitted the Affordability Study on March 10, 2016 (email correspondence with FEMA Congressional Affairs staff, March 10, 2016). However, Part 2 of the Affordability Study was available from the NAS website on December 11, 2015.

121 FEMA. An Affordability Framework for the National Flood Insurance Program, 2018.

122 See FEMA's website for the Community Status Book at https://www.fema.gov/national-flood-insurance-program-community-status-book.

of the floodplain management standards. Ultimately, if the community does not correct its cited deficiencies after given time periods described in regulations, the community will be suspended from the NFIP by FEMA.[123]

A community can also be involuntarily suspended from the NFIP for either

- failing to adopt an approved floodplain map and an approved set of floodplain management standards within the time periods required by regulations; or
- repealing or revising its floodplain management standards to a level below the minimum standards set forth in regulations.[124]

A suspended community may be reinstated to the NFIP once the relevant errors or deficiencies provoking the suspension have been resolved to meet FEMA's specification.

Communities that have been suspended or those communities that do not participate in the NFIP can face significant consequences. Primarily, members of these communities are not able to purchase primary flood insurance through the NFIP, which may result in significant uninsured property risk in that community. However, communities may elect not to participate in the NFIP because they have very little flood risk to begin with, given their particular geography or climate.

In addition, if a community does not participate in, or has been suspended from, the NFIP but has been previously mapped by FEMA for flood hazards, it is difficult for the community and policyholders to access other forms of federal assistance for areas in the floodplain.[125] For example, by law, no federal assistance may be provided for acquisition or construction purposes in an area that has been identified as having special

---

[123] For additional details on probation, see 44 C.F.R. 59.24(b) and (c).

[124] See 44 C.F.R. 59.24(a) and (d).

[125] For more on how individuals may receive assistance following a flood, see CRS Report R44808, *Federal Disaster Assistance: The National Flood Insurance Program and Other Federal Disaster Assistance Programs Available to Individuals and Households after a Flood*, by Diane P. Horn. For additional details on the effects of suspension, see FEMA's website at http://www.fema.gov/suspension.

flood hazards unless the property is covered by flood insurance.[126] Likewise, federally backed mortgages still require flood insurance for properties in the SFHA, so these property-owners would be required to obtain such insurance in the private market. A community is allowed to leave the NFIP at its will, but the potential consequences of that decision are similar to those if the community has been suspended.

FUNDING

The funding for the NFIP is primarily maintained in an authorized account called the National Flood Insurance Fund (NFIF).[127] Generally, the NFIP has been funded through three methods:

- receipts from the premiums of flood insurance policies, including fees and surcharges;
- direct annual appropriations for specific costs of the NFIP; and
- borrowing from the U.S. Treasury when the balance of the NFIF has been insufficient to pay the NFIP's obligations (e.g., insurance claims).

This section of the chapter briefly discusses each of these three methods of NFIP funding.

Premium Fees and Surcharges

As of February 2018, the written premium on approximately 5 million policies in force was $3.6 billion.[128] Included within the premiums are

---

[126] 42 U.S.C. §4012a(a).

[127] The NFIF is authorized by 42 U.S.C. §4017.

[128] The approximately $3.6 billion in premiums, for over 5 million policies, provides for $1.28 trillion in insured coverage across the nation. For full statistics, including breakdowns by state, see FEMA's website at http://bsa.nfipstat.fema.gov/reports/1011.htm.

several fees and surcharges on flood insurance policies mandated by law. First, the Federal Policy Fee (FPF) was authorized by Congress in 1990 and helps pay for the administrative expenses of the program, including floodplain mapping and some of the insurance operations.[129] The amount of the FPF is set by FEMA and can increase or decrease year to year. Since the April 2016 rating period, the FPF has been set at a flat rate of $50 for SFIPs, and $25 for PRPs. As of October 2017, the FPF is also $25 for contents-only policies.[130]

Second, a Reserve Fund assessment was authorized by Congress in BW-12 to establish and maintain a Reserve Fund to cover future claim and debt expenses, especially those from catastrophic disasters.[131] By law, FEMA is ultimately required to maintain a reserve ratio of 1% of the total loss exposure through the Reserve Fund assessment.[132] As of February 2018, the amount required for the Reserve Fund ratio was approximately $12.79 billion. However, FEMA is allowed to phase in the Reserve Fund assessment to obtain the ratio over time, with an intended target of not less than 7.5% of the 1% Reserve Fund ratio in each fiscal year (so, using February 2018 figures, not less than approximately $959 million each year). Since April 2016, using its discretion, FEMA has charged every NFIP policy a Reserve Fund assessment equal to 15% of the premium charged for both SFIPs and PRPs.[133] The Reserve Fund assessment has increased from its original status, in October 2013, of 5% on all SFIPs, and 0% on PRPs.[134]

In addition to the Reserve Fund assessment, all NFIP policies are also assessed a surcharge following the passage of HFIAA.[135] The amount of

---

129 42 U.S.C. §4014(a)(1)(B)(iii).

130 See FEMA, *Flood Insurance Manual, Rating Section*, Revised April 2018, p. RATE 16, https://www.fema.gov/media-library-data/1523307287100-4cf9726b2eb04c3471a3e9d37a58fa6a/05_rating_508_apr2018.pdf.

131 Section 100212 of P.L. 112-141, 126 Stat. 992, as codified at 42 U.S.C. §4017a.

132 42 U.S.C. §4017a(b).

133 See Federal Emergency Management Agency, *Flood Insurance Manual, Rating Section*, Revised April 2018, p. RATE 16, at https://www.fema.gov/media-library-data/1523307287100-4cf9726b2eb04c3471a3e9d37a58fa6a/ 05_rating_508_apr2018.pdf.

134 For additional information on the Reserve Fund, see FEMA, *Quarterly NFIP Reserve Fund Report*, June 15, 2016.

135 Section 8(a) of P.L. 113-89, 128 Stat. 1023.

the surcharge is dependent on the type of property being insured. For primary residences, the charge is $25; for all other properties, the charge is $250.[136] Revenues from the surcharge are deposited into the Reserve Fund.

Appropriations and Offsetting Receipts

Table 5 displays how Congress has appropriated and authorized offsetting receipts for the NFIP from FY2015 to FY2018. As provided for in law, all premiums from the sale of NFIP insurance are transferred to FEMA and deposited in the NFIF.[137] Congress then authorizes FEMA to withdraw funds from the NFIF, and use those funds for specified purposes needed to operate the NFIP. In addition to premiums, Congress has also provided annual appropriations to supplement floodplain mapping activities. In addition to the mix of discretionary and mandatory funding levels indicated in Table 5, which are set in appropriations legislation, fluctuating levels of mandatory spending occur in the NFIP in order to pay and adjust claims on affected NFIP policies.[138]

Borrowing from the U.S. Treasury, NFIP Debt

Congress has authorized FEMA to borrow no more than $30.425 billion from the U.S. Treasury in order to operate the NFIP. The authorization for this borrowing would be reduced to $1 billion after July 31, 2018, were the NFIP to be allowed to lapse.[139] In January 2017, the NFIP borrowed $1.6 billion due to losses in 2016 (the August 2016 Louisiana floods and Hurricane Matthew). On September 22, 2017, the

---

[136] For a description of the how the fee is applied to different policy types, see FEMA, *The HFIAA Surcharge Fact Sheet*, April 2015, at https://www.fema.gov/media-library/assets/documents/105569.

[137] 42 U.S.C. §4017(b).

[138] This mandatory spending is authorized by 42 U.S.C. §4017(d)(1). All other expenses using the NFIF must be authorized in appropriations acts, per 42 U.S.C. §4017(f).

[139] 42 U.S.C. §4016(a).

NFIP borrowed the remaining $5.825 billion from the Treasury to cover claims from Hurricane Harvey, Hurricane Irma, and Hurricane Maria, reaching the NFIP's authorized borrowing limit of $30.425 billion.[140] On October 26, 2017, Congress cancelled $16 billion of NFIP debt, making it possible for the program to pay claims for Hurricanes Harvey, Irma, and Maria.[141] This represents the first time that NFIP debt has been cancelled, although Congress appropriated funds between 1980 and 1985 to repay NFIP debt.[142] FEMA borrowed another $6.1 billion on November 9, 2017, to fund estimated 2017 losses, including those incurred by Hurricanes Harvey, Irma, and Maria and anticipated programmatic activities, bringing the debt up to $20.525 billion. The NFIP currently has $9.9 billion of remaining borrowing authority.[143]

The NFIP's debt to the U.S. Treasury cannot be tied directly to any single incident, as any insurance claim paid by the NFIP is in some way responsible for the existing debt of the NFIP (i.e., a dollar paid in claims, and therefore expended by the NFIP, following a minor flooding incident is no different than a dollar paid following a major hurricane). However, the NFIP was forced to borrow heavily to pay claims in the aftermath of two catastrophic flood seasons, the 2005 hurricane season (particularly Hurricanes Katrina, Rita, and Wilma) and Hurricane Sandy in 2012.[144] For example, following Hurricane Sandy, Congress passed P.L. 113-1 to increase the borrowing limit of the NFIP from $20.775 billion to the current $30.425 billion. Prior to Hurricane Katrina in 2005, the NFIP had generally been able to cover its costs, borrowing relatively small amounts from the U.S. Treasury to pay claims, and then repaying the loans with interest.

---

[140] Email correspondence from FEMA Congressional Affairs Staff, January 17, 2017 and September 22, 2017, respectively.

[141] P.L. 115-72, Title III, §308.

[142] Funds for "repayment under notes" were appropriated in P.L. 96-526, 94 Stat. 3053; P.L. 97-101, 95 Stat. 1425; P.L. 97-272, 96 Stat. 1169; P.L. 98-45, 97 Stat. 228; P.L. 98-371, 98 Stat. 1224; and P.L. 99-160, 99 Stat. 918.

[143] Email correspondence from FEMA Congressional Affairs staff, January 4, 2018.

[144] For accounting of the NFIP's premium revenues and claims/loss data, see FEMA's website for policy and claim statistics at https://www.fema.gov/policy-claim-statistics-flood-insurance.

Table 5. Budget Authority for the NFIP, FY2015-FY2018 (millions of dollars of budget authority, available for the fiscal year unless otherwise indicated)

Form of Budget Authority	Activity	FY2015, P.L. 114-4	FY2016, P.L. 114-113	FY2017 P.L. 115-31	FY2018 P.L. 115-141
Discretionary appropriations	Flood hazard mapping and risk analysis program[a] (available until expended)	$100.0	$190.0	$177.5	$262.5
Spending authority from offsetting collections, discretionary	Salaries and expenses associated with flood management and flood insurance operations	$23.8	$25.3	$13.4[b]	$13.57[b]
Spending authority from offsetting collections, mandatory (amount is shown as a limitation provided in appropriations legislation)	Floodplain management and flood mapping[c]	$155.5	$155.9	$168.4	$189.9
	Operating expenses	$136.0	$133.3	$147.0[b]	$1.165[b]
	Commissions and taxes of agents	$1,139.0	$1,123.0	$1,123.0	$1,123.0
	Interest on Treasury borrowings[d]	Such sums as necessary	Such sums as necessary	Such sums as necessary	Such sums as necessary
	Flood mitigation assistancee (available until expended)	$150.0	$175.0	$175.1	$175.0
	Flood Insurance Advocate[f]	$5.0	$5.0	$5.0	$5.0

Source: CRS analysis of P.L. 114-4, P.L. 114-113, P.L. 115-31, and P.L. 115-141.

[a] Generally, for necessary expenses, including administrative costs, related to the Risk MAP program authorized by 42 U.S.C. §§4101, 4101a, 4101b, 4101c, 4101d, and 4101e.

[b] The FY2017 and FY2018 budgets includes the amount of offsetting collections for *flood insurance operations* within the "operating expenses" activity, instead of the broader "salaries and expenses associated with flood management and flood insurance operations" activity as was done in P.L. 114-4 and P.L. 114-113.

[c] Offsetting receipts for "floodplain management and flood mapping" can generally been viewed as supplementing the discretionary appropriation for "flood hazard mapping and risk analysis program."

[d] The amount of interest paid on borrowed amounts for the U.S. Treasury fluctuates annually based on a number of factors, including the interest rate of the borrowing; the available funds for interest and principal payments after claims payments; the amount borrowed; how the debt is being serviced in loans, and fiscal decisions by FEMA to build the Reserve Fund as opposed to paying off principal and interest on the debt. FEMA reported interest payments of approximately $345.3 million for FY2016 (source: email correspondence from FEMA Congressional Affairs Staff, January 17, 2017).

[e] Flood mitigation assistance is authorized by 42 U.S.C. §4104c.

[f] The Flood Insurance Advocate is authorized by 42 U.S.C. §4033.

The NFIP's debt is conceptually owed by current and future participants in the NFIP, as the insurance program itself owes the debt to the Treasury and pays for accruing interest on that debt through the premium revenues of policyholders. For example, from FY2006 to FY2016 (i.e., since the NFIP borrowed funds following Hurricane Katrina), the NFIP has paid $2.82 billion in principal repayments and $3.83 billion in interest to service the debt through the premiums collected on insurance policies.[145]

Under its current authorization, the only means the NFIP has to pay off the debt is through the accrual of premium revenues in excess of outgoing claims, and from payments made out of the growing Reserve Fund. As required by law, FEMA submitted a report to Congress in 2013 on how the borrowed amount from the U.S. Treasury could be repaid within a 10-year period.[146] Whether or not FEMA will ultimately be able to pay off the debt is largely dependent on future insurance claims, namely if a catastrophic flooding incident such as Hurricanes Harvey, Sandy, or Katrina occurs again and with what frequency. However, using various predictions for both revenues (i.e., premiums) and losses (i.e., insurance claims), FEMA's report on debt repayment indicated even with the most optimistic scenario of future flooding it would take at least 13 years to repay the debt. In more realistic scenarios, the debt would not be paid off for at least 20 years, or it may increase considerably with future catastrophic incidents.[147] For example, in April 2017, CBO projected that the NFIP would have insufficient receipts to pay the expected claims and expenses over the 2018-2027 period and that FEMA would need to use about $1 billion of its then-current $5.825 billion borrowing authority to pay those expected claims.[148] Also in April 2017, FEMA updated some of the assumptions in the October 2015 NFIP Semi-Annual Debt Repayment Progress Report and estimated that at the end of 20 years, the NFIP's net debt would

---

[145] FEMA, *Semi-Annual NFIP Debt Repayment Progress Report*, October 2015, p. 16.

[146] See 42 U.S.C. §4016(d), as enacted by Section 100213(a) of P.L. 112-141 (BW-12).

[147] See FEMA, *Semi-Annual NFIP Debt Repayment Progress Report*, October 2015, p. 9.

[148] CBO, Preliminary Results from CBO's Analysis of the National Flood Insurance Program, at https://www.cbo.gov/ publication/52638.

increase by a further $9.4 billion.[149] No projections of the NFIP debt have yet been made that take account of the cancellation of $16 billion of NFIP debt, or the as yet unknown total claims of the 2017 hurricane season.

NFIP Purchase of Reinsurance

In HFIAA-14, Congress revised the authority of FEMA to secure reinsurance for the NFIP from the private reinsurance and capital markets.[150] In September 2016, FEMA secured its first placement of reinsurance for the NFIP, contracting for reinsurance cover which ran from September 19, 2016, through March 19, 2017, structured into two coverage layers. Under the first layer, the reinsurers would indemnify FEMA $1 million for flood claims losses that exceed $5 million. Under the second layer, the reinsurers would indemnify FEMA $1,000,000 when the total losses from a single flood event exceed $5.5 billion.[151] In January 2017, FEMA purchased $1.042 billion of insurance, to cover the period from January 1, 2017, to January 1, 2018, for a reinsurance premium of $150 million. Under this agreement, the reinsurance covered 26% of losses between $4 billion and $8 billion arising from a single flooding event. The purchase of private market reinsurance reduces the likelihood of FEMA needing to borrow from the Treasury to pay claims. However, since FEMA is withdrawing funds from the Reserve Fund to pay for this reinsurance, it subsequently increases the cost of insurance to policyholders. FEMA's modeling of the NFIP portfolio before the reinsurance purchase suggested that there was a 17.2% chance of losses from an event exceeding $4 billion in 2017.[152] FEMA has already paid over $8.6 billion in claims for Hurricane Harvey, triggering the full 2017 reinsurance.[153] In January 2018,

---

[149] Email correspondence from FEMA Congressional Affairs Staff, April 5, 2017. The revised assumptions used in making this projection were (1) accounting for the significant flooding in FY2016, including Baton Rouge flooding but excluding Hurricane Matthew; (2) revised premium increases for the final five years of the projection, resulting in lower premium and Reserve Fund projections; and (3) changes in the projected numbers of policies-in-force.

[150] See §10 of P.L. 113-89, 128 Stat. 1025, as codified at 42 U.S.C. §4081(e).

[151] See FEMA, *National Flood Insurance Program Reinsurance Initiative for 2016,* at https://www.fema.gov/national-flood-insurance-program-2016-reinsurance-initiative.

[152] See FEMA, *National Flood Insurance Program's Reinsurance Program for 2017*, at https://www.fema.gov/nfip-reinsurance-program.

[153] Email correspondence from FEMA Congressional Affairs Staff, April 20, 2018.

FEMA purchased $1.46 billion of insurance to cover the period from January 1, 2018, to January 1, 2019, for a reinsurance premium of $235 million. The agreement is structured to cover losses above $4 billion for a single flooding event, covering 18.6% of losses between $4 billion and $6 billion, and 54.3% of losses between $6 billion and $8 billion.[154] In April 2018, FEMA announced that it would seek to transfer additional NFIP risk to private markets through a reinsurance procurement in which the reinsurer acts as a transformer to transfer NFIP-insured flood risk through the issuance of a catastrophe bond, to be effective for a term of "likely" three years.[155]

Expiration of Certain NFIP Authorities

The statute for the NFIP does not contain a comprehensive expiration, termination, or sunset provision for the whole of the program. Rather, the NFIP has multiple different legal provisions that generally tie to the expiration of key components of the program. Unless reauthorized or amended by Congress, the following will occur after July 31, 2018:

- The authority to provide new flood insurance contracts will expire.[156] Flood insurance contracts entered into before the expiration would continue until the end of their policy term of one year.
- The authority for NFIP to borrow funds from the Treasury will be reduced from $30.425 billion to $1 billion.[157]

---

[154] FEMA, *National Flood Insurance Program (NFIP) Reinsurance Program*, at https://www.fema.gov/nfip-reinsurance-program.

[155] FEMA, *National Flood Insurance Program (NFIP) Reinsurance Program*, at https://www.fema.gov/nfip-reinsurance-program. For additional information on this, see CRS Insight IN10887, *The National Flood Insurance Program (NFIP), Reinsurance, and Catastrophe Bonds*, by Diane P. Horn and Baird Webel.

[156] 42 U.S.C. §4026.

[157] 42 U.S.C. §4016(a).

Other activities of the program would technically remain authorized following July 31, 2018, such as the issuance of FMA grants.[158] However, the expiration of the key authorities described above would have varied, generally serious effects on these remaining NFIP activities.

[158] See 42 U.S.C. §4104c and 42 U.S.C. §4104d.

In: Emergency Management
Editor: Stephanie Padbury
ISBN: 978-1-53614-101-6

Chapter 2

FEDERAL DISASTER ASSISTANCE: THE NATIONAL FLOOD INSURANCE PROGRAM AND OTHER FEDERAL DISASTER ASSISTANCE PROGRAMS AVAILABLE TO INDIVIDUALS AND HOUSEHOLDS AFTER A FLOOD*

Diane P. Horn
Flood Insurance and Emergency Management

ABSTRACT

After a flood, people are often uncertain if their eligibility for federal disaster assistance is linked in any way to whether or not they have flood insurance. Because much of the other disaster assistance that is available to individuals comes from the Federal Emergency Management Agency

* This is an edited, augmented and reformatted version of a Congressional Research Service report, R44808, dated February 13, 2018.

(FEMA), there may be confusion between possible claims provided through the National Flood Insurance Program (NFIP, which is also managed by FEMA), and other disaster assistance programs. This chapter provides an overview of the assistance available to individuals and households following a flood and provides links to more comprehensive guidance on both flood insurance and disaster assistance.

The National Flood Insurance Program (NFIP) is the main program intended to provide federal assistance to homeowners and renters recovering from flood losses. The maximum coverage for one- to four-family homes is $100,000 for contents and $250,000 for buildings coverage. In addition to NFIP claims payments to policyholders, homeowners and renters may also access a number of other federal programs aimed at mitigating the impact on individuals and households. The principal FEMA program to offer assistance to individuals and families is the Individual and Households Program (IHP). The total of all IHP assistance to one household cannot exceed $34,000. IHP recipients whose homes are located in a Special Flood Hazard Area, are in a community participating in the NFIP, and who receive assistance for repair, replacement, permanent housing construction, and/or personal property as a result of a flood-related disaster must obtain and maintain flood insurance as a condition of accepting disaster assistance.

The Small Business Administration (SBA) Disaster Loan Program provides direct loans to businesses, nonprofit organizations, homeowners, and renters to repair or replace property destroyed in a federally declared disaster. A Personal Property Loan provides a creditworthy homeowner or renter in a declared disaster area with up to $40,000 to repair or replace personal property owned by the survivor, while Real Property Loans provide creditworthy homeowners with up to $200,000 to repair or restore the homeowners' primary residence to its predisaster condition. Recipients of SBA loans must carry flood insurance for the life of the loan.

In some instances that are perceived as catastrophic events, Congress has provided additional resources to states and local governments through the Department of Housing and Urban Development's Community Development Block Grant Disaster Recovery Program (CDBG-DR); however, the CDBG-DR program is not automatically triggered by a disaster. Due to the block grant nature of the program, local and state officials exercise a great deal of discretion in determining which combination of eligible activities to employ. This allows communities to use CDBG-DR funds to meet disaster-related needs, including short-term disaster relief, mitigation activities, and long-term recovery activities. HUD does not provide CDBG-DR funding directly to individuals; however, individuals and families may benefit from a number of the eligible activities for which CDBG-DR funds can be used. Unless the individual state requires the purchase of flood insurance, the recipients of CDBG-DR grants are exempt from the requirement to purchase flood insurance.

The NFIP's authorization expires on March 23, 2018. If the NFIP is not reauthorized and is allowed to lapse, the authority to provide new flood insurance contracts will expire. If the NFIP were to lapse, the unavailability of NFIP insurance could have an impact on other programs such as IHP, SBA disaster loans, and CDBG-DR. Separately from or in conjunction with NFIP reauthorization, the impacts on such other programs may be useful to consider. In particular, Congress may consider revising the requirements for flood insurance in the Flood Disaster Protection Act.

INTRODUCTION

After a flood event, people are often uncertain about what types of federal assistance they are eligible for, and if their eligibility for federal disaster assistance is linked in any way to whether or not they have flood insurance. Because much of the other disaster assistance that is available to individuals comes from the Federal Emergency Management Agency (FEMA), there may be confusion between possible claims provided through the National Flood Insurance Program (NFIP, which is also managed by FEMA) and other disaster assistance programs. This chapter provides an overview of the assistance available to individuals and households following a flood and provides links to more comprehensive guidance on both flood insurance and disaster assistance.

The National Flood Insurance Program (NFIP) is the main program intended to provide federal assistance to homeowners and renters recovering from flood losses. The NFIP was established by the National Flood Insurance Act of 1968 (NFIA, 42 U.S.C. §4001 et seq.). The general purpose of the NFIP is both to offer primary flood insurance to properties with significant flood risk, and to reduce flood risk through the adoption of floodplain management standards. As of November 2017, the NFIP had about 5 million flood insurance policies providing over $1.27 trillion in coverage.[1] Nationally, as of December 2017, about 22,302 communities in 56 states and jurisdictions participated in the NFIP.[2] According to FEMA, the program saves the nation an estimated $1.87 billion annually in flood losses avoided because of the NFIP's building and floodplain management regulations.[3]

In addition to NFIP claims payments to policyholders, homeowners and renters may also access a number of other federal programs aimed at mitigating the impact on individuals and households. As described below, Congress established certain limits on the availability of federal disaster assistance dependent on whether individuals had flood insurance, as set out by the Flood Disaster Protection Act of 1973 (FDPA, P.L. 93-234, 87 Stat 985). In addition to requiring property owners with federally backed mortgages to have flood insurance,[4] the FDPA restricts access to federal financial assistance for *acquisition or construction purposes*[5] for use in any area that has been identified as having special flood hazards (i.e., what is known as a Special Flood Hazard Area, or SFHA), and in which the sale of flood insurance has been made available under the NFIP, unless covered

---

[1] Statistics on the NFIP policy and claims are available from FEMA's website at https://www.fema.gov/policy-claim-statistics-flood-insurance.

[2] Based on FEMA's map inventory, 98.8% of the U.S. population is mapped with an existing flood map. Over 88% of the population lives in a community that has received a modernized product. Email correspondence from FEMA Congressional Affairs staff, April 20, 2017. Detailed information about which communities participate and where is available from the Community Status Book, found on FEMA's website at https://www.fema.gov/national-flood-insurance-program-community-status-book.

[3] Ibid.

[4] The mandatory purchase requirement is discussed in this report in the section entitled Primary Flood Insurance through the NFIP.

[5] Financial assistance for acquisition or construction purposes is defined at 42 U.S.C §4003(4) and means

any form of financial assistance which is intended in whole or in part for the acquisition, construction, reconstruction, repair, or improvement of any publicly or privately owned building or mobile home, and for any machinery, equipment, fixtures, and furnishings contained or to be contained therein, and shall include the purchase or subsidization of mortgages or mortgage loans but shall exclude assistance pursuant to the Disaster Relief and Emergency Assistance Act [42 U.S.C. 5121 et seq.] (other than assistance under such Act in connection with a flood).

Financial assistance is also separately defined at 42 U.S.C. §4003(3) and means "any form of loan, grant, guaranty, insurance, payment, rebate, subsidy, disaster assistance loan or grant, or any other form of direct or indirect Federal assistance, other than general or special revenue sharing or formula grants made to States."

by flood insurance (i.e., the community participates in the NFIP).[6] The Robert T. Stafford Disaster Relief and Emergency Assistance Act (P.L. 93-288, as amended, Stafford Act) also sets restrictions to flood disaster assistance, stating that notwithstanding any other provision of law, no federal disaster relief assistance made available in a flood disaster area may be used to make a payment (including any loan assistance payment) to a person for repair, replacement, or flood disaster assistance that was conditional on the person first having obtained flood insurance under applicable federal law and subsequently having failed to obtain and maintain flood insurance as required under applicable federal law on such property.[7]

FEDERAL ASSISTANCE AVAILABLE TO INDIVIDUALS AND FAMILIES

The National Flood Insurance Program (NFIP)

Availability of Flood Insurance through the NFIP

A core design feature of the NFIP is that communities[8] are not required to participate in the program by any law or regulation. Rather, communities in the United States voluntarily participate in the NFIP generally as a means of securing access to the primary flood insurance offered by the NFIP. Essentially, the NFIP is structured so that the availability of primary flood insurance through the NFIP is tied to the adoption and enforcement of floodplain management standards by participating communities. FEMA is only allowed to provide flood insurance to "those States or areas (or subdivisions thereof)" where "adequate land use and control measures" have been adopted that "are

---

[6] 42 U.S.C §4012a(a).

[7] 42 U.S.C. §5154a(a).

[8] 44 C.F.R. 59.1 defines community as "any State or area or political subdivision thereof, or any Indian tribe or authorized tribal organization, or Alaska Native village or authorized native organization, which has authority to adopt and enforce flood plain management regulations for the areas within its jurisdiction."

consistent with the comprehensive criteria for land management and use developed" by the NFIP.[9] Thus, communities that participate in the NFIP, and therefore whose residents may access the NFIP's primary flood insurance, also must adopt, through local or state laws, minimum floodplain management standards that are described in FEMA regulations. Communities are required to adopt these minimum floodplain management standards in order to participate in the NFIP.[10] FEMA has set forth the minimum standards it requires for participation in the NFIP in federal regulations.[11] Though the standards appear in federal regulations, the standards only have the force of law because they are adopted and enforced by a state or local government.

A community that has been found failing to enforce the floodplain management standards may be placed on probation and ultimately suspended from the NFIP. However, a community cannot be removed from the NFIP because of increased or excess flood insurance claims and losses. Rather, probation and suspension only occur if the community fails to uphold its obligations related to floodplain management.[12] Communities that have been suspended or those communities that do not participate in the NFIP can face significant consequences. Most importantly, residents of these communities are not able to purchase primary flood insurance through the NFIP, which may result in significant uninsured property risk in that community. In addition, if a community does not participate in, or has been suspended from, the NFIP but has been previously mapped by FEMA for flood hazards, it is difficult for the community and policyholders to access some other forms of federal assistance for areas in the floodplain.[13]

---

[9] 42 U.S.C §4012(c)(2).

[10] 42 U.S.C. 4022(a)(1).

[11] See 44 C.F.R. Part 60, particularly 44 C.F.R. 60.3.

[12] The conditions under which a community can be placed on probation or suspended from the NFIP are described in detail in CRS Report R44593, *Introduction to the National Flood Insurance Program (NFIP)*, pages 20-21.

[13] For additional details on the effects of suspension, see FEMA's website at http://www.fema.gov/suspension.

For example, by law, no federal assistance may be provided for acquisition or construction purposes in an area that has been identified as having special flood hazards unless the property is covered by flood insurance.[14] Likewise, as federally backed mortgages require flood insurance for properties in the Special Flood Hazard Area (SFHA),[15] these property owners would be required to obtain such insurance in the private market. A community is allowed to leave the NFIP at its will, but the potential consequences of that decision are similar to those if the community has been suspended.

The Coastal Barrier Resources System and Flood Insurance

In some other special cases NFIP flood insurance may not be available; for example, no new NFIP flood insurance coverage[16] may be provided for new construction or structures with substantial improvements which are located on any coastal barrier within the John H. Chafee Coastal Barrier Resources System (CBRS).[17] The CBRS contains two types of units: System units[18] and Otherwise Protected Areas (OPAs).[19] System units are generally comprised of private lands that were relatively undeveloped at the time of their designation within the CBRS. The boundaries of system units are generally intended to follow geomorphic, development, or cultural features. OPAs are generally comprised of lands held by a qualified organization primarily for wildlife refuge, sanctuary, recreational,

---

[14] 42 U.S.C. §4012a(a).

[15] A Special Flood Hazard Area (SFHA) is defined by FEMA as an area having special flood, mudflow or flood-related erosion hazards and shown on a Flood Hazard Boundary Map or a Flood Insurance Rate Map. See the following section on Primary Flood Insurance through the NFIP for a more detailed description of SFHAs.

[16] 42 U.S.C. §4028.

[17] The Coastal Barrier Resources Act (P.L. 97-348), as amended, designates undeveloped coastal barriers and adjacent areas, where most federal spending that would support additional development is prohibited. 42 U.S.C. §4028(a) prohibits the issuance of new flood insurance coverage in the CBRS. More details about the CBRA can be found in CRS Report RL34339, *Coastal Zone Management: Background and Reauthorization Issues*. For additional information about flood insurance and the Coastal Barrier Resources System, see 44 C.F.R. Part 71, particularly 71.3, and the NFIP Flood Insurance Manual, revised October 2017, Section CBRS 1, which can be accessed at https://www.fema.gov/ media-library-data/1503240263089-30b35cc754f462fe2c15d857519a71ec/19_cbrs_508_oct2017.pdf.

[18] 16 U.S.C. §3502.

[19] §12(2) of P.L. 101-591.

or natural resource conservation purposes. The boundaries of OPAs are generally intended to coincide with the boundaries of conservation or recreation areas such as state parks and national wildlife refuges.[20]

Primary Flood Insurance through the NFIP

FEMA develops, in coordination with participating communities, flood maps called Flood Insurance Rate Maps (FIRMs) that depict the community's floodplain and flood risk. A key aspect of this flood mapping is the identification of the Special Flood Hazard Area (SFHA). The SFHA is intended to distinguish the flood risk zones that have a chance of flooding during a "1 in 100 year flood" or greater frequency. This means that properties have a risk of 1% *or greater* of flooding every year if located in a SFHA. In a community that participates or has participated in the NFIP, owners of properties in the mapped SFHA are required to purchase flood insurance as a condition of receiving a federally backed mortgage. By law and regulation, federal agencies, federally regulated lending institutions, and government-sponsored enterprises must require these property owners to purchase flood insurance as a condition of any mortgage that these entities make, guarantee, or purchase.[21] Property owners falling under this mandate may purchase flood insurance through the NFIP or through a private company, so long as the private flood insurance "provides flood insurance coverage which is at least as broad as the coverage" of the NFIP, among other conditions.[22] Not all mortgages in the SFHA are affected by this mandatory purchase requirement. For example, a personal mortgage loan between two private parties (such as between family members), or a mortgage issued by a private mortgage company that is *not* then sold on the secondary market to a bank or entity like Fannie Mae, *may* not require flood insurance. Even if they are not

[20] See U.S. Fish and Wildlife Service information on the Coastal Barrier Resources Act at https://www.fws.gov/ ecological-services/habitat-conservation/cbra/Act/index.html.
[21] 42 U.S.C. §4012a.
[22] 42 U.S.C §4012a(b).

technically required to mandate flood insurance by federal law, the issuing party may still require it as a means of financially securing the property.

Preferred Risk Policies

Flood insurance is optional for properties outside the SFHA regardless of whether they have a federally backed mortgage. However, as there is still a risk of flooding outside the SFHA, residents of NFIP participating communities with property located outside the SFHA may voluntarily purchase a lower-cost Preferred Risk Policy (PRP). Rates for PRPs are lower than for properties in SFHAs in accordance with their lower risk profile; however, unlike properties in the SFHA, an individual may be denied a PRP if there is significant loss history for the property.[23] FEMA encourages the purchase of PRPs both to reduce the financial flood risk of a broader group of individuals, and to expand the number of NFIP policyholders, thus improving the fiscal soundness of the NFIP portfolio.

Standard Flood Insurance Policies

FEMA has considerable discretion under the law to craft the details of the flood insurance policies it sells through the NFIP.[24] Currently, there are three policies that the NFIP uses to sell primary flood insurance—the Dwelling, the General Property, and the Residential Condominium Building Association policy (RCBAP) forms. Collectively, these Standard Flood Insurance Policies (SFIPs) appear in regulations, and coverage qualifications are generally equivalent.[25]

Within the SFIPs sold by the NFIP, there are numerous policy exclusions that are often not understood by policyholders. For example,

---

[23] For additional information on PRPs and restrictions, see FEMA, *Flood Insurance Manual, Preferred Risk Policy Section*, Revised October 2017, at https://www.fema.gov/media-library-data/1503239443436-30b35cc754f462fe2c15d857519a71ec/09_prp_508_oct2017.pdf.

[24] 42 U.S.C. §4013(a).

[25] See 44 C.F.R. Part 61, Appendix A. Copies of the policy forms are also available on FEMA's website at https://www.fema.gov/national-flood-insurance-program/standard-flood-insurance-policy-forms.

SFIPs have limited coverage of basements or crawlspaces.[26] Because SFIP coverage limits are often less than the value of a structure or the value of the property's contents, policyholders can obtain excess flood insurance to cover losses beyond the coverage limit. However, such excess coverage is not sold by the NFIP, and can only be purchased through the private insurance market.

NFIP Coverage Limits: Single Family Dwellings

The maximum coverage limits available from the NFIP are determined by occupancy type and location in or out of a SFHA (see Table 1). The NFIP defines three main occupancy types: Single Family Dwelling,[27] Other Residential Building,[28] and Business Building.[29] The distinction between residential and nonresidential is determined by the percentage of the floor space which is devoted to commercial uses.

Policyholders are able to elect coverage for both their building property and separate coverage for contents. Renters may obtain a

---

[26] For a basic guide to coverage under the SFIP, see FEMA, *Summary of Coverage*, FEMA F-679, November 2012, at http://www.fema.gov/media-library-data/20130726-1620-20490-4648/f_679_summaryofcoverage_11_2012.pdf.

[27] FEMA's definition of a Single-Family Dwelling is either (1) a residential single-family building in which the total floor area devoted to nonresidential uses is less than 50% of the building's total floor area, or (2) a single-family residential unit within a 2-4 family building, other residential building, business, or nonresidential building, in which commercial uses within the unit are limited to less than 50% of the unit's total floor area. See FEMA, *Flood Insurance Manual, Definitions Section*, October 2017, p. DEF 9, https://www.fema.gov/media-library-data/1503240583997-30b35cc754f462fe2c15d857519a71ec/23_definitions_508_oct2017.pdf.

[28] FEMA defines an Other Residential Building as a residential building that is designed for use as a residential space for 5 or more families, or a mixed-use building in which the total floor space devoted to nonresidential uses is less than 25% of the total floor area within the building. See FEMA, *Flood Insurance Manual, Definitions Section*, October 2017, p. DEF 7, https://www.fema.gov/media-library-data/1503240583997-30b35cc754f462fe2c15d857519a71ec/23_definitions_508_oct2017.pdf.

[29] FEMA defines a Business Building as a building in which the named insured is a commercial enterprise primarily carried out to generate income and the coverage is for (1) a building designed as a nonhabitational building; (2) a mixed-use building in which the total floor area devoted to residential uses is (a) 50% or less of the total floor area within the building if the residential building is a single family property or (b) 75% or less of the total floor area within the building for all other residential properties; or (3) A building designed for use as office or retail space, wholesale space, hospitality space, or for similar uses. See FEMA, *Flood Insurance Manual, Definitions Section*, October 2017, p. DEF1, https://www.fema.gov/media-library-data/1503240583997-30b35cc754f462fe2c15d857519a71ec/ 23_definitions_508_oct2017.pdf.

contents-only coverage. The maximum coverage with either a SFIP or a PRP for single-family dwellings (which also includes single-family residential units within a 2-4 family building) is $100,000 for contents and up to $250,000 for buildings coverage.

NFIP Coverage Limits: Other Residential Buildings and Condominiums

For Other Residential Buildings which are not covered by a Residential Condominium Building Association Policy (RCBAP), the maximum available coverage limit is $250,000 for building coverage and $100,000 for contents coverage. These limits apply to all single condominium units and all other buildings not in a condominium form of ownership, including cooperatives, timeshares, apartment buildings, dormitories, and assisted-living facilities. The Other Residential Building category also includes hotels, motels, tourist homes, and rooming houses which have more than four units where the normal guest occupancy is six months or more.[30] An individual dwelling unit in a condominium not covered by the RCBAP may be separately insured under the Dwelling form, in the name of the unit owner, up to the limits of insurance coverage for a single-family dwelling (up to $100,000 for contents, up to $250,000 for building elements). A residential condominium association may purchase insurance coverage under the RCBAP on behalf of the association and the unit owners to cover the building and, if desired, coverage of commonly owned contents. In this case, the maximum limit for the building is the replacement cost, or the total number of units multiplied by $250,000, whichever is less.[31]

---

30 FEMA, *Flood Insurance Manual, General Rules Section*, October 2017, p. GR6, https://www.fema.gov/media-library-data/1503238892702-30b35cc754f462fe2c15d857519a71ec/03_general_rules_508_oct2017.pdf.

31 FEMA, *Flood Insurance Manual, Condominiums Section*, October 2017, p. CONDO 1 https://www.fema.gov/media-library-data/1503239181440-30b35cc754f462fe2c15d857519a71ec/06_condo_508_oct2017.pdf.

NFIP Claims

While FEMA provides the overarching management and oversight of the NFIP, the bulk of the day-to-day operation of the NFIP, including claims made after a flood, is handled by private companies. This arrangement between the NFIP and private industry is authorized by statute and guided by regulation.[32] There are two different arrangements that FEMA has established with private industry. The first is the Direct Servicing Agent (DSA), which operates as a private contractor on behalf of FEMA for individuals seeking to purchase flood insurance policies directly from the NFIP. The second arrangement is called the Write-Your-Own (WYO) Program, where private insurance companies are paid to directly write and service the policies themselves. With either the DSA or WYO Program, the NFIP retains the actual financial risk of paying claims for the policy (i.e., underwrites the policy), and the policy terms and premiums are the same.

Following a flood, NFIP policyholders are advised to contact their agent or insurance company to file a claim. This is either one of the WYO companies or the DSA, who will provide an adjuster to assist the policyholder to make a claim. The adjuster will normally take measurements and photographs to assess the damage, provide an estimate of the covered flood damage, and provide a proof of loss form for the policyholder to sign. The estimate, proof of loss form, and other supporting documentation provide the Proof of Loss,[33] which is required before the claim can be paid. The complete Proof of Loss, signed and sworn to by the policyholder, along with documentation to support the amount requested initially and any requests for additional payment, must be sent to the DSA or WYO Company within 60 days after the date of loss or within the

---

[32] See primarily 42 U.S.C. §4081 and §4018, and 44 C.F.R. Part 62.

[33] The Proof of Loss is the policyholder's statement of the amount of money being requested, signed to and sworn by the policyholder with documentation to support the amount requested. The Proof of Loss is not the claim. The claim is the policyholder's declaration that they are entitled to be paid for a covered loss under the terms of the NFIP flood insurance policy. For additional information on the claims procedure, see FEMA, *Flood Insurance Manual, Claims Section*, October 2017, at https://www.fema.gov/media-library-data/1503240095424- 30b35cc754f462fe2c15d857519a71ec/15_claims_508_oct2017.pdf.

extension. In severe events, FEMA may authorize an extension of the 60-day period to send the proof of loss to the NFIP adjuster. For example, following the Louisiana floods in August 2016, policyholders were given 270 days to submit their proof of loss,[34] and after Hurricane Sandy in 2012, FEMA authorized multiple extensions, which ultimately gave policyholders a full two years to submit their proofs of loss.[35] Policyholders have been given one year from the date of loss to file claims for Hurricane Harvey,[36] Hurricane Irma,[37] and Hurricane Maria.[38]

Advance Payments from the NFIP

In some circumstances, FEMA has authorized a conditional waiver to allow insurers (either WYO companies or the NFIP Direct Servicing Agent) to make advance payments to policyholders.[39] These advance payments have been made in two ways.

1. Pre-inspection: Once a policyholder provides notice of loss, an insurer may offer an advance payment for building and/or personal

---

[34] Federal Emergency Management Agency, Third Extension of the Deadline for Sending Proofs of Loss in the State of Louisiana for Claims Related to the Mid-Summer Severe Storms Commencing on August 9, 2016 through August 31, 2016, W-17002, February 3, 2017, at http://bsa.nfipstat.fema.gov/wyobull/2017/w-17002.pdf.

[35] Federal Emergency Management Agency, Further Extension of the Time to File a Standard Flood Insurance Policy ("SFIP") Proof of Loss for National Flood Insurance Program ("NFIP") Flood Claims Arising Out of Meteorological Event Sandy, W-14017, April 28, 2014, at http://bsa.nfipstat.fema.gov/wyobull/2014/w-14017.pdf.

[36] Federal Emergency Management Agency, Activation of NFIP Catastrophic Event Enhanced Claim Payment Process Hurricane Harvey, W-17030, September 3, 2017, at https://nfip-iservice.com/Stakeholder/pdf/bulletin/w-17030.pdf.

[37] Federal Emergency Management Agency, Activation of NFIP Catastrophic Event Enhanced Claim Payment Process and Proof of Loss Extension for Hurricane Irma, W-17040, September 12, 2017, at https://nfip-iservice.com/ Stakeholder/pdf/bulletin/w-17040.pdf.

[38] Federal Emergency Management Agency, Activation of NFIP Catastrophic Event Enhanced Claim Payment Process and Proof of Loss Extension for Hurricane Maria, W-17057, September 28, 2017, at https://nfip-iservice.com/ Stakeholder/pdf/bulletin/w-17057.pdf.

[39] For example, FEMA Memorandum W-16053 (August 17, 2016) and FEMA Memorandum W-17039 (September 12, 2017) both authorized a conditional waiver of the Standard Flood Insurance Policy, Section VII. General Conditions, Paragraphs (J) and (M) and Section VIII, General Conditions, Paragraphs (J) and (M) of the Residential Condominium Building Association Policy to allow advance payments. See https://bsa.nfipstat.fema.gov/ wyobull/2016/w-16053.pdf and https://nfip-iservice.com/Stakeholder/pdf/bulletin/w-17039.pdf.

property damages up to a total of $5000 after confirming coverage and validating that the insured property has flooded. The insurer may offer a total advance payment of up to $20,000 if the policyholder provides photographs depicting flood damage to the covered property and either documentation such as receipts verifying out-of-pocket expenses related to the repair or replacement of covered property, or a contractor's itemized damage estimate.

2. Payment for Significant Damage: If the insurer receives a general contractor's estimate of necessary repairs to the insured property and a flood insurance inspector retained by the insurer has inspected the insured property, an advance payment may be authorized, not to exceed 50% of the general contractor's estimate of necessary repairs. In both cases, the advance payment cannot be used for Additional Living Expenses,[40] and after the claim is settled, the insurer will reduce the final payment by the amount of the advance. If the loss is determined not to be a covered loss, or if the advance payment exceeds the amount of the actual covered loss, the insured will be required to repay the advanced payment (or portion thereof).[41]

FEMA Assistance to Individuals and Families

When the President declares a major disaster under the Stafford Act, the President's action contains a designation for the types of assistance FEMA will provide for victims of that disaster. Individual Assistance (IA) may be provided for certain declared disaster areas, but not others. IA provides help to individuals and families and can include several programs,

---

[40] Additional Living Expenses coverage in a standard homeowner's policy reimburses the insured for the cost of maintaining a comparable standard of living following a covered loss that exceeds the insured's normal expenses prior to the loss.

[41] See https://bsa.nfipstat.fema.gov/wyobull/2016/w-16053.pdf and https://nfip-iservice.com/Stakeholder/pdf/bulletin/w-17039.pdf.

depending on whether the governor of the affected state or the tribal leader has requested that specific help.[42]

Individuals and Household Program

The principal IA program to offer assistance to individuals and families is the Individual and Households Program (IHP).[43] The IHP has two broad categories that assist families and individuals who have been impacted by disaster damage: housing assistance and other needs assistance. The total of all assistance to one household cannot exceed $34,000.[44] Each household is considered a single applicant regardless of the number of household members.[45] In all instances, this help is intended to supplement, but not substitute for, existing insurance coverage. If applicants have insurance, any assistance provided by FEMA for losses covered by insurance should be considered an advance and must be repaid to FEMA upon receipt of an insurance payment settlement.

Under IHP, housing costs can be assumed by the federal government for up to 18 months.[46] Unlike other Stafford Act programs, housing is not cost shared with the state, but is all federal assistance. The types of housing assistance include monetary assistance, direct assistance, repairs, and replacement. Monetary Assistance is provided to individuals and families to address the costs to rent temporary housing while repairs are being made to their predisaster primary residence or while they transition to permanent housing. The amount of rental assistance available is calibrated to include housing and utility costs in the affected area. Direct Assistance provides temporary housing units to disaster survivors that are purchased or leased

---

[42] For additional information on FEMA Individual Assistance, see CRS Report R44619, *FEMA Disaster Housing: The Individuals and Households Program—Implementation and Potential Issues for Congress*.

[43] 42 U.S.C. 5174 §408.

[44] This amount is adjusted annually in accordance with the Consumer Price Index. The latest adjustment was posted in Federal Emergency Management Agency, "Notice of Maxiumum Amount of Assistance Under the Individuals and Households Program," 82(196) *Federal Register* 47568, October 12, 2017.

[45] 44 C.F.R. 206.101(e)(2).

[46] 42 U.S.C. 5174 §408(c)(1)(B)(iii). This period can be extended by the President for "extraordinary circumstances" in the "public interest."

by the federal government. The IHP cost per household limit does not apply to Direct Assistance. The use of Direct Assistance, such as mobile homes and trailers, is rare and is generally considered a last resort to be employed only when other housing options are not available in the immediate disaster area.

The great majority of disaster housing help comes in the form of temporary rental assistance and repairs to a home which is not insured in order to make it habitable. This repair assistance does not include new utilities or improvements to the home, but does make the home conform to current, applicable building codes in the affected area. The goal is to repair the home to a safe and sanitary living or functioning condition. FEMA will not pay to return a home to its condition before the disaster. The IHP can also make a contribution through Replacement Assistance toward the replacement of an uninsured or underinsured owner-occupied residence that was damaged in a disaster event. However, the $34,000 limit means that it is unlikely that IHP awards will pay for the full replacement of a residence. IHP recipients whose homes are located in a Special Flood Hazard Area and are in a community participating in the NFIP and who receive assistance for repair, replacement, permanent housing construction, and/or personal property as a result of a flood-related disaster must obtain and maintain flood insurance as a condition of accepting disaster assistance.[47]

Other Needs Assistance

Funding for individuals and families is also available through the Other Needs Assistance (ONA) program. Unlike housing, ONA is cost shared with the state government on a 75% federal, 25% state basis. The ONA program is intended to address specific needs created by the disaster's impact. Assistance under ONA can include clothing, furniture, funeral expenses, child care, emergency medical and/or dental help, and a

[47] 44 C.F.R. 206.110 (k)(3).

range of other needs.[48] Flood insurance may be required on insurable items (personal property) if they are to be located in a SFHA.[49]

Group Flood Insurance

One unique form of assistance that is eligible for ONA funding is the Group Flood Insurance Policy (GFIP) offered by FEMA to IHP recipients after major flood disasters.[50] The GFIP is a temporary mechanism for recipients of IHP funds—generally low-income persons—to acquire flood insurance following a flood loss. This assistance allows households time to recover from the disaster and be in a better position to buy flood insurance after the expiration of the GFIP.[51] The premium for the GFIP is a flat fee of $600 per insured household, which is paid from ONA funds. A special deductible of $200 (applicable separately to any buildings loss and any contents loss) applies to insured flood damage losses sustained by the insured property in the course of any subsequent flooding event during the term of the GFIP. The policy covers the maximum amount of IHP awards (currently $34,000) and begins 60 days after the date of the applicable disaster declaration. The term of the policy is 36 months and is nonrenewable. FEMA is to provide a certificate of coverage to each household and contact the household 60 days before the policy is due to expire. The expiration notification from FEMA is designed, in part, to encourage policyholders to contact a local insurance agent or producer or a private insurance company selling NFIP policies under the WYO program, and advise them as to the amount of coverage the policyholder must maintain in order not to jeopardize their eligibility for future disaster assistance.

---

[48] 44 C.F.R. 206.119-120. For a detailed listing of ONA assistance, see FEMA Individual and Households Program Fact Sheet, June 1, 2016, available at https://www.fema.gov/media-library/assets/documents/24945.

[49] FEMA, *What is FEMA's Individual Assistance Program*, June 16, 2016, https://www.fema.gov/media-library-data/1461689021638-cfcfd7f6c263635802fa7a76a19e00ea/FS001_What_is_Individual_Assistance_508.pdf.

[50] 44 C.F.R. §61.17.

[51] 44 C.F.R. 206.119(d).

Other Individual Assistance

Other programs which may be available to individuals and families from FEMA include Crisis Counseling for disaster victims,[52] Disaster Legal Assistance,[53] and Disaster Unemployment Assistance (DUA).[54] The latter is available if there is a significant number unemployed due to the disaster who do not qualify for the regular state unemployment program. DUA is federally funded through FEMA but is administered by the Department of Labor and State Unemployment Compensation agencies. These programs do not require flood insurance.

Transitional Shelter Assistance

FEMA also uses Public Assistance grants to provide short-term shelter for disaster survivors who are unable to return to their homes.[55] For instance, Transitional Shelter Assistance (TSA) provides short-term sheltering assistance to disaster survivors who have a continuing need for shelter after congregate shelters[56] have closed. The TSA initiative is intended to provide short-term lodging for eligible disaster survivors whose communities are either uninhabitable or inaccessible due to disaster-related damages. FEMA considers applicants' eligibility for TSA as part of the IHP application process.[57] TSA is authorized by the Stafford Act[58] and is subject to Public Assistance regulations on state cost share. However, TSA is administered through FEMA's Individual Assistance program, which is better prepared to collect information on individuals than Public

---

[52] Stafford Act §416, 42 U.S.C. §5177.

[53] 42 U.S.C. §5182.

[54] Stafford Act §410, 20 C.F.R. §625.

[55] For additional information on Public Assistance, see CRS Report R43990, *FEMA's Public Assistance Grant Program: Background and Considerations for Congress*.

[56] FEMA defines a congregate shelter as any public or private facility that provides refuge to evacuees but that normally serves a nonrefuge function, such as schools, churches, or stadiums. See Federal Emergency Management Agency, *Mass Sheltering and Housing Assistance*, Recovery Strategy RS-2006-1, July 24, 2006, https://www.fema.gov/media-library-data/20130726-1824-25045-9144/mass_sheltering_and_housing_assistance.pdf.

[57] U.S. Government Accountability Office, *Hurricane Sandy: FEMA Has Improved Disaster Aid Verification but Could Act to Further Limit Improper Assistance*, GAO-15-15, December 2014, pp. 10-11, http://www.gao.gov/ products/GAO-15-15.

[58] §403, 42 U.S.C. §5170b and §502, 42 U.S.C. §5192.

Assistance.[59] For those who are eligible for TSA support, FEMA has the authority to fund the use of hotels or motels as transitional shelters, through direct payment to the participating hotels or motels.

Sheltering and Temporary Essential Power Program

In the aftermath of Hurricane Sandy in 2012, FEMA designed the Sheltering and Temporary Essential Power (STEP) Pilot Program to work with state, local, and tribal governments to carry out certain essential measures to help restore power, heat, and hot water to residential properties, allowing residents to remain in or return to their homes while permanent repairs were completed. The STEP program home repairs were not accomplished under the authority of the IHP program, but instead were funded under Section 403 of the Stafford Act,[60] which covers essential assistance; in other words, the grants were provided to eligible Public Assistance Grant Program applicants, who in turn used the funding for essential repairs to private residences. The STEP program was intended to reduce the demand for other FEMA-provided shelter, such as congregate shelters or TSA. STEP consisted of three elements: residential electrical meter repairs, shelter essential measures, and rapid temporary exterior repairs. FEMA delivered the STEP program through direct federal assistance; reimbursement of applicants[61] who perform, or contract for the performance of, authorized emergency protective measures; or a combination of the two. Individual residential property owners were not eligible to apply directly to STEP. Eligible costs under the STEP pilot program were limited to $10,000 per residential unit, including equipment, materials, labor, and any associated inspection fees that were necessary to

---

[59] U.S. Government Accountability Office, *Hurricane Sandy: FEMA Has Improved Disaster Aid Verification but Could Act to Further Limit Improper Assistance*, GAO-15-15, December 2014, p. 11, http://www.gao.gov/products/ GAO-15-15.

[60] 42 U.S.C. §5170b.

[61] Applicants can be state, local or tribal governments.

accomplish work that was eligible under STEP.[62] STEP assistance was made available to residents in New York, New Jersey, and Connecticut.[63]

The STEP pilot program shared some commonality with IHP, as it provided basic repairs to disaster-affected residences. However, STEP was intended only to make homes habitable for the short term and did not provide any assistance beyond basic repairs to homes, whereas IHP provides financial assistance for home repair or replacement, rental expenses, damaged personal property, and other needs. In New York City, the New York City Rapid Repair Program augmented STEP to allow for repairs to multifamily dwellings and more permanent repairs needed to meet building codes.[64] As STEP funding was not provided to individuals, support from STEP did not affect the determination of any IA grant from FEMA. Households which benefitted from STEP support were not required to have or purchase flood insurance.[65]

FEMA instituted a similar program to STEP, called Shelter at Home, to address housing needs following the floods in Louisiana in August 2016 (FEMA-4277-DR-LA).[66] Eligible costs under the Shelter at Home program were limited to single-family owner-occupied properties for repairs up to a maximum value of $15,000. As with the STEP program, Shelter at Home assistance did not affect the determination of any IA grant from FEMA and there was no flood insurance requirement for recipients of Shelter at Home assistance.[67]

STEP programs have been instituted in Texas following Hurricane Harvey (FEMA-4332-DR-TX), in Florida following Hurricane Irma

---

[62] FEMA Recovery Program Guidance: *Sheltering and Temporary Essential Power (STEP) Pilot Program*, November 6, 2012, https://www.fema.gov/media-library-data/20130726-1858-25045-8258/step_pilot_program_final_111612.pdf.

[63] U.S. Government Accountability Office, *Hurricane Sandy: FEMA Has Improved Disaster Aid Verification but Could Act to Further Limit Improper Assistance*, GAO-15-15, December 2014, p. 12, http://www.gao.gov/products/ GAO-15-15.

[64] U.S. Government Accountability Office, *Hurricane Sandy: FEMA Has Improved Disaster Aid Verification but Could Act to Further Limit Improper Assistance*, GAO-15-15, December 2014, p.11, http://www.gao.gov/products/ GAO-15-15.

[65] U.S. Government Accountability Office, *Hurricane Sandy: FEMA Has Improved Disaster Aid Verification but Could Act to Further Limit Improper Assistance*, GAO-15-15, December 2014, p. 11, http://www.gao.gov/products/ GAO-15-15.

[66] https://www.fema.gov/disaster/4277.

[67] For further information about Shelter at Home, see https://shelterathome.la.gov/.

(FEMA-4337-DR-FL), in Puerto Rico following Hurricane Irma (FEMA-4336-DR-PR) and Hurricane Maria (FEMA-4339-DR-PR), and in the U.S. Virgin Islands following Hurricane Irma (FEMA-4335-DR-VI) and Hurricane Maria (FEMA-4340-DR-VI). All of these STEP programs are available only for disaster damaged single-family owner-occupied primary residential properties, including duplexes and townhomes. The Texas program is known as Partial Repair and Essential Power for Sheltering (PREPS).[68] All of the STEP programs for the 2017 hurricanes capped the cost of repairs at $20,000. As with previous STEP programs, STEP assistance for the 2017 hurricanes did not affect a FEMA IHP applicant's eligibility for repair, replacement, or permanent or semi-permanent housing construction assistance, if approved, under Section 408 of the Stafford Act,[69] and there was no flood insurance requirement for recipients other than providing documentation of flood insurance or certification of no flood insurance.[70]

The Coastal Barrier Resources System and Housing Assistance

The Coastal Barrier Resources Act (P.L. 97-348, as amended through P.L. 107-136) restricts housing assistance provided under Section 408 of the Stafford Act. Housing assistance, including repair, replacement, and semi-permanent or permanent construction, is not authorized to those who live in the Coastal Barrier Resources System (CBRS) or in an Otherwise Protected Area (OPA).[71] However, essential assistance under Section 403

---

[68] Federal Emergency Management Agency, Partial Repair and Essential Power for Sheltering (PREPS) Pilot Program for FEMA-4332-DR-TX , November 20, 2017, memo provided by FEMA Congressional Affairs staff, November 20, 2017.

[69] 42 U.S.C. 5174 §408.

[70] Federal Emergency Management Agency, Sheltering and Temporary Essential Power (STEP) Pilot Program for FEMA-4337-DR-FL, Final Policy Approval, September 30, 2017; Federal Emergency Management Agency, Sheltering and Temporary Essential Power (STEP) Pilot Program for FEMA-4336-DR-PR and FEMA-4339-DR-PR, October 29, 2017; Federal Emergency Management Agency, Sheltering and Temporary Essential Power (STEP) Pilot Program for FEMA-4335-DR-VI and FEMA-4340-DR-VI, October 27, 2017, memos provided by FEMA Congressional Affairs staff, November 13, 2017; and Federal Emergency Management Agency, Partial Repair and Essential Power for Sheltering (PREPS) Pilot Program for FEMA-4332-DR-TX, November 20, 2017, memo provided by FEMA Congressional Affairs staff, November 20, 2017.

[71] 16 U.S.C. §3504.

of the Stafford Act is allowed in the CBRS and in OPAs.[72] In addition, although FEMA discourages some types of sheltering in CBRA or OPA areas, none of the STEP policies exclude implementation of STEP in CBRAs/OPAs. STEP is funded as an emergency protective measure under 403 and is, therefore, exempt from most of the CBRA/OPA restrictions.[73]

SBA Disaster Loan Program

The Small Business Administration (SBA)[74] Disaster Loan Program provides direct loans to businesses, nonprofit organizations, homeowners, and renters to repair or replace property damaged or destroyed in a federally declared disaster. SBA Disaster Loans include Home and Personal Property Disaster Loans, Business Physical Disaster Loans, and Economic Injury Disaster Loans. About 83% of direct disaster loans are awarded to individuals and households rather than small businesses.[75] The program generally offers disaster loans at a fixed rate that have loan maturities of up to 30 years. The SBA Disaster Loan Program can be put into effect by five types of declarations: two types of Presidential Declarations as authorized by the Stafford Act, and three types of SBA declarations.[76] The type of declaration determines what types of loans are made available, but has no bearing on the loan terms or loan caps.

SBA Personal Property Loans

A Personal Property Loan provides a creditworthy homeowner or renter in a declared disaster area with up to $40,000 to repair or replace personal property owned by the survivor, while Real Property Loans

---

[72] 16 U.S.C. §3505(6)(E).

[73] Email correspondence from FEMA Congressional Affairs staff, November 13, 2017.

[74] For further information about the Small Business Administration, see CRS Report R41309, *The SBA Disaster Loan Program: Overview and Possible Issues for Congress* and CRS Report R44412, *SBA Disaster Loan Program: Frequently Asked Questions*.

[75] 13 C.F.R. §123.3. Further information on SBA Disaster Loans is available in CRS Report R44412, *SBA Disaster Loan Program: Frequently Asked Questions*.

[76] See 13 C.F.R. §123.3(a)(3), 13 C.F.R. 123.3(a)(4), 13 C.F.R. §123.3(a)(5), and 15 U.S.C. §632(k). A summary of the types of declarations that trigger SBA loans is available in CRS Report R44412, *SBA Disaster Loan Program: Frequently Asked Questions*.

provide creditworthy homeowners with up to $200,000 to repair or restore the homeowners' primary residence to its predisaster condition. Only uninsured or otherwise uncompensated disaster losses are eligible.[77] The amount that SBA will lend depends on the cost of repairing or replacing the home and/or personal property (minus insurance settlements or grant assistance). The loans may not be used to upgrade a home or build additions to a home unless the upgrade or addition is required by city or county building codes.[78] However, a homeowner can borrow additional funds beyond the cost of repair to cover the costs of mitigation intended to protect their property against future damage; however, mitigation funds may not exceed 20% of the disaster damage, as verified by SBA, to a maximum of $200,000 for home loans.[79]

Disaster loans may be used in conjunction with other types of assistance, including insurance, but only to the extent that there is no duplication of benefit. The Stafford Act requires federal agencies providing disaster assistance to ensure that businesses and individuals do not receive disaster assistance for losses for which they have already been compensated.[80] SBA regulations prohibit applicants from receiving a home disaster loan if their damaged property can be repaired or replaced with the proceeds of insurance, gifts, or other compensation.[81] These amounts must either be deducted from the amount of the claimed losses or, if received after SBA has approved and disbursed a loan, as principal payments on their loans.[82]

SBA Flood Insurance Requirements

Recipients of SBA loans must carry flood insurance for the life of the loan.[83] Because the National Flood Insurance Reform Act of 1994, among

---

[77] 13 C.F.R. §123.105(a)(1) and 13 C.F.R. §123.105(a)(2).
[78] 13 C.F.R. §123.105(a)(2).
[79] 13 C.F.R. §123.107.
[80] Stafford Act §312, 42 U.S.C. §5177.
[81] For a discussion of rules prohibiting the Duplication of Benefits, see CRS Report R44553, *SBA and CDBG-DR Duplication of Benefits in the Administration of Disaster Assistance: Background, Policy Issues, and Options for Congress*.
[82] 13 C.F.R. 123.101(c).
[83] 13 C.F.R. §120.170.

other things, made flood insurance requirements directly applicable to agencies that provide government loans, such as the SBA,[84] such agencies cannot subsidize, insure, or guarantee any loan if the property securing the loan is in a SFHA of a community not participating in the NFIP. In addition, the SBA will deem an applicant ineligible for a home disaster loan if the SBA determines that the applicant assumed the risk by not maintaining flood insurance as required by an earlier SBA disaster loan when the current loss is also due to flood.[85] SBA loans are not available for properties in the Coastal Barrier Resources System (CBRS) or in an Otherwise Protected Area (OPA).[86]

Assistance Which May Be Available to Individuals and Families after a Flood

In addition to NFIP and other federal assistance that may be automatically triggered with a disaster declaration, individuals may also access federal assistance that is not automatically triggered by such declarations. In some instances that are perceived as catastrophic events, Congress has provided additional resources to states and local governments through the Department of Housing and Urban Development's Community Development Block Grant Disaster Recovery Program (CDBG-DR). The CDBG-DR program is not automatically triggered by a disaster. Instead, Congress has occasionally addressed unmet disaster needs by providing supplemental disaster-related appropriations for the CDBG-DR program. Consequently, CDBG-DR is not provided for all major disasters declared under the Stafford Act, but only provided at the discretion of Congress.

---

[84] 42 U.S.C. §4012a(b)(2).
[85] 13 C.F.R. §123.101(d).
[86] 13 C.F.R. §120.175.

HUD Community Development Block Grant Disaster Recovery Program

The CDBG-DR program is designed to help communities and neighborhoods that otherwise might not recover after a disaster due to limited resources. Eligible grantees include states, units of local government, Indian tribes, and other areas designated by the President as disaster areas. Generally, CDBG-DR grantees must use at least 70% of the funds for activities that principally benefit low- and moderate-income (LMI) persons or areas. Due to the block grant nature of the program, local and state officials exercise a great deal of discretion in determining which combination of eligible activities to employ. This allows communities to use CDBG-DR funds to meet disaster-related needs, including short-term disaster relief, mitigation activities, and long-term recovery activities.[87] Communities and states are also free to amend and revise their annual CDBG plans in response to flood in order to address imminent threats to the health and safety of residents. Communities, at the discretion of Congress, may be reimbursed by CDBG-DR funds for use of their regular CDBG allocation. This results from explicit language included in some appropriations bills.[88]

HUD does not provide CDBG-DR funding directly to individuals; however, individuals and families may benefit from a number of the eligible activities for which CDBG-DR funds can be used. Examples of these activities include buyouts of damaged properties in a floodplain and relocating residents to safer areas, relocation payments for people and businesses displaced by a disaster, rehabilitation of homes and buildings damaged by the disaster, and homeownership activities such as down payment assistance, interest rate subsidies, and loan guarantees for disaster victims.[89]

---

[87] CRS Report RL33330, *Community Development Block Grant Funds in Disaster Relief and Recovery*.

[88] For a further discussion of CDBG allocations, see CRS Report RL33330, *Community Development Block Grant Funds in Disaster Relief and Recovery*.

[89] Department of Housing and Urban Development, *CDBG-DR Eligibility Requirements*, at https://www.hudexchange.info/programs/cdbg-dr/cdbg-dr-eligibility-requirements/.

Examples of CDBG-DR Assistance after Major Floods

Some of the biggest flood-related CDBG-DR grants have been made available in the aftermath of hurricanes. For example, following the 2005 hurricane season (particularly Hurricanes Katrina, Rita, and Wilma), Congress provided $19.7 billion in CDBG-DR funding to five states (Alabama, Florida, Louisiana, Mississippi, and Texas). Louisiana and Mississippi were the two states hardest hit by the 2005 hurricanes and received $13.4 billion and $5.5 billion, respectively. Both states devoted the majority of the CDBG-DR funding to housing recovery, with particular support for homeowners. Louisiana state officials used $3 billion of CDBG-DR funding for the Road Home program. This program defined a homeowner's potential grant amount as the estimated cost to repair or rebuild the home, minus any insurance payments and FEMA assistance for structural repairs, up to a maximum of $150,000. This potential amount was then reduced by 30% for homeowners who did not carry insurance (homeowners insurance and also flood insurance if the property was in a SFHA and the community participated in the NFIP). The state also reduced the award to remove Duplication of Benefits and unpaid taxes owed to the state.[90] Homeowners could choose to repair their property or to relocate (with only 60% of the potential grant amount available if they chose to relocate out of Louisiana). Mississippi established a similar program called the Mississippi Homeowner Grant Program. The maximum assistance in the Mississippi program considered the applicant's income relative to the area median income, and whether or not the property was located in a SFHA. Assistance in Mississippi was also conditional on homeowners' willingness to attach covenants to the property that created commitments to maintain flood insurance, rebuild and repair in accordance with applicable building codes and local ordinances, and elevate the property in accordance with the most recent FEMA guidelines.[91]

---

[90] Jonathan Spader and Jennifer Turnham, "CDBG Disaster Recovery Assistance and Homeowners' Rebuilding Outcomes Following Hurricanes Katrina and Rita," *Housing Policy Debate*, vol. 24, no. 1 (2014), pp. 213-237.

[91] HUD, *Recovery Snapshot: Mississippi Homeowners Grant Program*, March 31, 2006, 2 pp. https://www.hudexchange.info/onecpd/assets/File/CDBG-DR-Case-Study-Mississippi-Homeowner-Compensation-Grants.pdf.

After Hurricane Sandy, which made landfall in October 2012, $16 billion ($15.8 billion after the sequester) was awarded in CDBG-DR funding to New York City and the states of Connecticut, Maryland, New Jersey, New York, and Rhode Island. This funding has been used in a variety of ways which provide funding directly to individuals and families, including housing rehabilitation and replacement, resilient rebuilding (for example, elevation of a property), property acquisition, and coastal resiliency measures. The State of New Jersey used CDBG-DR funding after Hurricane Sandy to make grants of up to $10,000 through its Homeowner Resettlement Program which homeowners could use toward addressing anything that would provide an incentive to remain in their community. One of the eligible expenses for which this money could be used was to pay flood insurance premiums.[92] The Resettlement Program was only available to homeowners in the nine most impacted counties whose property was their primary residence which they owned and occupied at the time of the storm, and whose residence sustained a FEMA-verified loss of $8,000 or greater or more than one foot of flooding on the first floor. Before the funds were released, the homeowner was required to sign a promissory note to reside in the county for three years; otherwise, they would have to repay the funds to the state. Initially 60% of the funds were reserved for LMI families in accordance with HUD income guidelines. The disbursement of the funds was evaluated in the order in which the requests were received. Following the 2017 Hurricanes (Harvey, Irma, and Maria), Congress initially made $7.39 billion available in CDBG-DR funding.[93] Table 1 provides summary information on available federal assistance from an NFIP insurance policy, FEMA Individual and Households Program, and SBA disaster loans. This information is provided dependent on the status of the individual's property in the Special Flood Hazard Area, the community's participation in the NFIP, and whether that individual has the required insurance coverage.

---

[92] New Jersey Department of Community Affairs. Community Development Block Grant Disaster Recovery Action Plan. March 2013. http://www.state.nj.us/dca/announcements/pdf/CDBG-DisasterRecoveryActionPlan.pdf. See pages 4-7 to 4-9 for a detailed description of the New Jersey Homeowner Resettlement program.

[93] P.L. 115-56.

Table 1. Available Funding Based on NFIP Status

NFIP Status	Occupancy Type	Maximum Available from NFIP Policy	Maximum Available from FEMAIHP If in Stafford Act Declared Disaster Area	Maximum Available from SBA Disaster Loan After a Declaration
In SFHA with flood insurance, community participates in NFIP	1-4 family structures Condominium units Other residential structures, RCBAP policies, business structures	Building: $250,000 Contents; $100,000 Building: $250,000 Contents; $100,000 Building: $500,000 Contents: $100,000	Individual Assistance: up to limit of $34,000; IA reduced by amount of NFIP proceeds received	SBA Personal Property Loan: up to $40,000 SBA Real Property Loan: up to $200,000 Recipients of SBA loans must carry flood insurance for the life of the loan
In SFHA without flood insurance when required to do so, community participates in NFIP	All occupancy types	Nothing	Individual Assistance: up to limit of $34,000; must obtain and maintain flood insurance as a condition of accepting housing assistance	Nothing
Not in SFHA with Preferred Risk Policy, community participates in NFIP	1-4 family structures Other residential structures	Building: $250,000 Contents: $100,000 Building: $250,000 Contents: $100,000	Individual Assistance: up to limit of $34,000	SBA Personal Property Loan: up to $40,000 SBA Real Property Loan: up to $200,000

NFIP Status	Occupancy Type	Maximum Available from NFIP Policy	Maximum Available from FEMAIHP If in Stafford Act Declared Disaster Area	Maximum Available from SBA Disaster Loan After a Declaration
	Business structures	Buildings: $500,000 Contents: $500,000		Recipients of SBA loans must carry flood insurance for the life of the loan
Not in SFHA, without flood insurance, community participates in NFIP	All occupancy types	Nothing	Individual Assistance: up to limit of $34,000	SBA Personal Property Loan: up to $40,000 SBA Real Property Loan: up to $200,000 Recipients of SBA loans must carry flood insurance for the life of the loan
In Coastal Barrier Resources System (CBRS)or Otherwise Protected Areas (OPA)	All occupancy types	Nothing—properties in CBRS cannot buy flood insurance	Individual Assistance: essential assistance under Section 403 of the Stafford Act may be available by exception; housing assistance is not available	Nothing—SBA may not make or guarantee any loan within the CBRS
Not in SFHA, community does not participate in NFIP or withdraws from NFIP or is suspended from NFIP	All occupancy types	Nothing—residents of communities which do not participate in the NFIP cannot buy NFIP flood insurance	Individual Assistance: up to limit of $34,000	SBA Personal Property Loan: up to $40,000 SBA Real Property Loan: up to $200,000 Recipients of SBA loans must carry flood insurance for the life of the loan

Table 1. (Continued)

NFIP Status	Occupancy Type	Maximum Available from NFIP Policy	Maximum Available from FEMAIHP If in Stafford Act Declared Disaster Area	Maximum Available from SBA Disaster Loan After a Declaration
In SFHA, community does not participate in NFIP or withdraws from NFIP or is suspended from NFIP	All occupancy types	Residents of suspended communities cannot purchase or renew NFIP flood insurance policies, but policies in force at the time of suspension remain in force for the policy term	Federal assistance is not available for the acquisition or construction of buildings in identified SHFAs, if the community includes an area which has been identified as a SFHA for at least one year; essential assistance under Section 403 of the Stafford Act may be available	SBA loans are not available

Source: Information on NFIP: 42 U.S.C. §4001 et seq., 42 U.S.C. §4106(a), and FEMA, *Flood Insurance Manual, Rating Section*, Revised October 2016, at https://www.fema.gov/media-library/assets/documents/127010.

Information on IHP: 42 U.S.C. §5174(h) and FEMA, *Federal Register* 81, October 1, 2016, page 40431. Information on SBA: 13 C.F.R. §123.105(a)(2) and 13 C.F.R. §123.120.170. Information on CBRS: 13 C.F.R. §123.120.175 and 13 C.F.R. §123.120.17.

SUMMARY AND CONSIDERATIONS FOR CONGRESS

Unequal Amount of Assistance Available through Flood Insurance and Disaster Assistance

There is often a perception that flood victims get large amounts of disaster aid and that this may act as a moral hazard which discourages people from purchasing flood insurance and/or adopting risk reduction measures.[94] In fact, the amount of FEMA Individual Assistance after a flood is relatively limited; many of the headline reports of federal aid refer to public assistance or assistance which is not available to all individuals. Flood insurance may be preferable for governments, since policyholders cover a portion of the costs through premium payments. Flood insurance may also be better for individuals, as uninsured households fare less well than those with insurance. Generally, residents must live in a federally declared disaster area to be eligible to receive disaster assistance or disaster loans; in contrast, NFIP policyholders may claim for damages from any flood, regardless of whether the flood was a declared disaster under federal law. Although the maximum coverage available under the NFIP may not be sufficient to cover all flood-related losses, an individual or family will generally be able to get more from NFIP insurance than from disaster assistance. For example, homeowners are able to get up to $350,000 for buildings and contents together and renters are able to get up to $100,000, compared to a maximum of $34,000 per household from FEMA disaster assistance. In addition, most disaster victims do not receive the maximum amount available under FEMA disaster assistance. For example, after the South Carolina floods in October 2015, although less than 5% of households in the counties receiving IHP had NFIP flood insurance, the average IHP payment was $3,199.[95] In contrast, the average NFIP claim for the South Carolina floods was $34,911.[96] Neither flood insurance payments nor Individual Assistance payments need to be paid back; however, the amount available through IA will be reduced by the amount of NFIP proceeds received for the same damages. In addition, a household without flood insurance will only be able to get IA under a limited set of circumstances. An individual or household may be able to borrow up to

---

[94] For example, see Carolyn Kousky and Leonard Shabman, *The Realities of Federal Disaster Aid: The Case of Floods*, Resources for the Future, Issue Brief 12-02, Washington, DC, April 2012, p. 1, http://www.rff.org/files/sharepoint/WorkImages/Download/RFF-IB-12-02.pdf.

[95] Figures calculated by CRS from data available at https://www.fema.gov/disaster/4241 and from data provided in email correspondence from FEMA Congressional Affairs, February 8, 2017.

[96] Federal Emergency Management Agency, Significant Flood Events, at https://www.fema.gov/significant-flood-events.

$240,000 through a combination of SBA Personal Property and Real Property Loans; however, these loans must be paid back with interest. In order to receive SBA loans, recipients must carry flood insurance for the life of the loan. The amount of this assistance is subject to Duplication of Benefits and sequence protocols.

The nexus between flood insurance and disaster assistance reveals conflicting aims: the desire to encourage individuals and households to purchase flood insurance versus the humanitarian objective of providing assistance following a disaster so that people do not suffer unnecessarily. This is fundamentally a moral hazard question. A fully individualistic approach to flood risk would advocate that individuals and households should purchase adequate flood insurance to protect themselves against a major flood disaster. A more egalitarian approach could take the view that individuals do not necessarily have a free choice on whether or not they purchase flood insurance (for example, residents of a community which does not participate in the NFIP) and that some individuals and households may not be able to afford sufficient coverage. In this case, the policy question is whether or not the federal government should step in after a disaster even if the victims do not have flood insurance. Congress may allow these conflicting policies to continue and to retain the current difference between the amounts available from flood insurance and disaster assistance, or may consider proposals that would alter the balance between assistance and insurance.

Availability of Alternative Funding

A significant amount of money is available to individuals and households through funding sources which are not available in all disasters, particularly through CDBG-DR. For example, homeowners in Louisiana and Mississippi could get grants of up to $150,000 after the 2005 hurricane season, as could homeowners in New Jersey after Hurricane Sandy. Rather than being funded through the annual appropriations process, statutory authority for CDBG-DR funding is provided via supplemental appropriations which are awarded by Congress only in extraordinary circumstances that have resulted in significant unmet needs for long-term recovery. Because Congress has so far opted to use special appropriations to deliver CDBG-DR funds rather than establishing a permanent CDBG disaster recovery program, HUD cannot create permanent regulations and guidance of how states should implement CDBG-DR grants.[97] This may lead to delay and/or confusion in implementation, and may also subject program design decision to local politics.[98] Unless the individual state required homeowners to purchase flood insurance, the recipients of CDBG-DR grants are exempt from the requirement to purchase flood insurance.[99] Given the unequal use of CDBG-DR, there may be an issue of equity in disaster housing assistance, where some disaster events receive an expanded form of federal assistance to meet housing needs in the form of CDBG, which can create different tiers of disaster housing assistance for declared disasters.[100]

Communities Which Do Not Participate in the NFIP

Congress may consider whether they agree that individual households should be disadvantaged by a decision made by their community, over which an individual is likely to have very little control. Individuals who live in a community which has not joined, withdraws from, or which has been suspended from the NFIP, are not allowed to buy or renew NFIP flood insurance and are barred from applying for SBA loans. These

---

[97] Recently HUD has developed a set of waivers and alternatives to reflect lessons learned from CDBG-DR assistance.

[98] See Jonathan Spader and Jennifer Turnham, "CDBG Disaster Recovery Assistance and Homeowners' Rebuilding Outcomes Following Hurricanes Katrina and Rita," *Housing Policy Debate*, vol. 24, no. 1 (2014), pp. 213-237. They argued that both Louisiana and Mississippi allocated a disproportionate share of CDBG-DR funds after the 2005 hurricanes to homeowner properties compared with rental properties.

[99] 24 C.F.R. §58.6(a)(3).

[100] See CRS Report R44619, *FEMA Disaster Housing: The Individuals and Households Program—Implementation and Potential Issues for Congress* for a fuller discussion of this issue.

households may wish to buy NFIP flood insurance but are not allowed to do so, because of their community's NFIP status (or nonstatus).

ACKNOWLEDGMENTS

Eugene Boyd (G&F), Jared Brown (G&F), and Bruce R. Lindsay (G&F) were consulted in the preparation of this chapter.

In: Emergency Management
Editor: Stephanie Padbury

ISBN: 978-1-53614-101-6

Chapter 3

THE FIRST RESPONDER NETWORK (FIRSTNET) AND NEXT-GENERATION COMMUNICATIONS FOR PUBLIC SAFETY: ISSUES FOR CONGRESS*

Jill C. Gallagher
Telecommunications Policy

ABSTRACT

During the events of September 11, 2001 (9/11), first responders could not communicate with each other. Some radios did not work in the high-rise World Trade Center; radio channels were overloaded by the large number of responders trying to communicate; and public safety radio systems operated on various frequencies and were not interoperable. There were also non-technical issues. Officials struggled to coordinate the multi-agency response, and to maintain command and control of the numerous agencies and responders.

* This is an edited, augmented and reformatted version of a Congressional Research Service report, R45179, dated April 27, 2018.

The 9/11 Commission called for the "expedited and increased assignment of radio spectrum for public safety purposes." Increased spectrum would allow public safety agencies to accommodate an increasing number of users; support interoperability solutions (e.g., shared channels); and leverage new technologies (e.g., live video streams) to enhance response.

In 2012, Congress acted on the recommendation of the 9/11 Commission. In Title VI of the Middle Class Tax Relief and Job Creation Act of 2012 (P.L. 112-96), Congress authorized the Federal Communications Commission (FCC) to allocate additional spectrum for public safety use; established the First Responder Network Authority (FirstNet) and authorized it to enter into a public-private partnership to build a nationwide public safety broadband network; and, provided $7 billion out of revenues from spectrum auctions to build the network.

February 22, 2017, marked five years since the act was signed into law. FirstNet has made progress in implementing the provisions in the act. In March 2017, FirstNet awarded a 25-year, $6.5 billion contract to AT&T to build and maintain the nationwide network for public safety. FirstNet provided AT&T with 20 megahertz (MHz) of broadband spectrum, which AT&T can monetize for public safety and non-public safety use. AT&T is providing FirstNet access to its infrastructure, valued at $180 billion, and $40 billion to maintain and improve the network.

In September 2017, FirstNet/AT&T presented states with plans detailing how the network would be deployed in each state. Governors could opt to have AT&T deploy the network (i.e., opt in), or have the state assume responsibility for the deployment (i.e., opt out). By January 2018, all 50 states and 6 territories opted in. This was viewed as a victory for FirstNet, AT&T, and public safety stakeholders who had long advocated for a nationwide network for public safety.

However, challenges remain. While governors allowed FirstNet/ AT&T to deploy the network in their states, there is no requirement for state and local public safety agencies to use the network. FirstNet/AT&T must attract users to the network to ensure the network is self-sustaining, as required under the act. FirstNet set adoption targets and steep penalties that AT&T must pay if targets are not met. AT&T has offered specialized features and services (e.g., priority access to the network, support during disasters) to attract users to the network. However, Verizon has offered similar services to entice users to its network which may affect FirstNet/AT&T's enrollment efforts. There are other factors affecting enrollment. Some public safety agencies have expressed reluctance to join the FirstNet network, citing uncertainties with the resiliency, reliability, and security of the network, coverage, and cost. Other agencies have expressed an unwillingness to join until FirstNet can provide mission critical voice features—essential features that responders have on their radios and use during emergencies—that will not be available from FirstNet until 2019. Attracting users to the network will be challenging for FirstNet/AT&T, but necessary to meet the requirements in the law and achieve the intent of the act.

Congress may continue its oversight of FirstNet to ensure the FirstNet network is meeting public safety needs (e.g., security, reliability, and resiliency), requirements in the law are met, and the network is deployed as intended. Congress may monitor subscribership to ensure the network will be self-sustaining, as required in the act, and that the intent of the law is achieved.

INTRODUCTION

The First Responder Network Authority (FirstNet) is a federal agency created by Congress in the Middle Class Tax Relief and Job Creation Act of 2012 (P.L. 112-96)—to oversee the building, deployment, and operation of a new nationwide public safety broadband network. In P.L. 112-96 (the act), Congress created FirstNet as an independent authority, within the National Telecommunications and Information Administrations (NTIA), under the Department of Commerce. FirstNet is governed by a Board comprised of federal, private sector, and public safety representatives, which reviews and approves the major decisions of the agency.[1]

Congress authorized $7 billion be allocated out of revenues from spectrum auctions that were authorized in the act. Many in Congress recognized that $7 billion would not be enough to deploy a nationwide network for public safety.[2] As a result, Congress authorized FirstNet to enter into a public-private partnership to leverage commercial infrastructure and services, in exchange for access to FirstNet assets, including $7 billion in funding and 20 megahertz (MHz) of spectrum.

The act is complex. It requires the establishment of a new federal agency and the appointment of a Board of Directors. It requires the

[1] The Board is comprised of the DHS Secretary, U.S. Attorney General, Office of Management and Budget (OMB) Director, and 12 representatives with various expertise appointed by the Secretary of Commerce (P.L. 112-96 § 6204).

[2] National Public Safety Telecommunications Council, "Senators Worried about Cost of Public Safety Broadband Network," February 16, 2011, http://www.npstc.org/documents/Senators WorriedAboutCostPSBBNet110216.pdf.

development and deployment of a new nationwide broadband network for public safety users. The act requires the auction of spectrum to fund FirstNet—a complex and lengthy process that concluded in 2015. The act requires FirstNet to build the network to industry standards, and the development of technical requirements, some of which were still under development by the international standards development organization, 3GPP.[3] Further, the act requires consultation with public safety users; governors; and federal, state, local, tribal, and territorial entities to coordinate the deployment of the network. Each mandated task took time but was used to inform a request for proposal (RFP) which was released in January 2016, and awarded in March 2017.

In March 2017, FirstNet awarded the contract to build the network to AT&T. In May 2017, AT&T offered public safety agencies priority and preemption services on its nationwide network to attract users to the network.[4] In September, FirstNet/AT&T released state plans detailing the initial (five-year) deployment of the network in the state. By January 2018, all states and territories had opted in to the network (i.e., allowed AT&T to deploy the network in their state).[5]

AT&T is deploying the core network, and plans to deploy the radio access network (e.g., cell site equipment, antennas) in each state. AT&T is also recruiting users to the network to ensure adoption targets are met. FirstNet plans to oversee the network deployment, ensuring: network elements are completed on time, integrated, and interoperable; public safety needs, including coverage, resiliency, and security are met; development and testing of mission critical voice capabilities and devices remain on schedule; user fees and devices are affordable; and that users subscribe to the network, so the network will be self-sustaining.

---

[3] 3GPP is a (private) standards development organization (SDO) that develops specifications that define 3GPP technologies. For information on 3GPP, see Lair, Yannick and Mayer, Georg, "Mission Critical Service in 3GPP," *3GPP*, June 20, 2017. See http://www.3gpp.org/news-events/3gpp-news/1875-mc_services.

[4] Priority access is a specialized offering that moves first responders to the front of the communications line during periods when communication networks may be difficult to access or congested. Preemption shifts non-emergency traffic to another line to free up space for public safety officials to communicate with others.

[5] For some activities, FirstNet and AT&T acted jointly, which will be depicted in this report as "FirstNet/AT&T."

Background on FirstNet Spectrum

Since the 1920s, first responders have used land mobile radios (LMR), such as radios in vehicles and handheld devices, to communicate during day-to-day operations and emergencies. Radios allow responders to transmit messages quickly over the airwaves to coordinate and communicate during response. To transmit messages over the airwaves, public safety agencies must obtain a license from the FCC which regulates the use of the radio spectrum in the United States.[6]

The FCC assigns frequencies to non-federal users, including state and local public safety agencies. Pursuant to the Communications Act of 1934 (47 U.S.C. § 151 et. seq.), the FCC designated certain segments of spectrum for specific uses. The FCC set aside a swathe of spectrum for public safety; however, demand soon exceeded the amount of spectrum allocated. As a result, the FCC began assigning various frequencies across the radio spectrum to public safety agencies. While the increased assignment of spectrum gave public safety agencies the spectrum they needed to communicate, public safety agencies were now operating on various frequency bands, scattered across the spectrum, using band-specific equipment that was not interoperable with equipment made to operate in other bands. Public safety advocates called for the allocation of additional spectrum and solutions that would enable agencies to interoperate.[7] In the Balanced Budget Act of 1997 (P.L. 105-33), Congress allocated 24 MHz to public safety. This required the transition of broadcasters off the spectrum, but provided an indefinite timeline for their

[6] In general, spectrum is the range of radio frequencies used for radio, television, and other electromagnetic communications. Specifically, the radio frequency spectrum is the part of the natural spectrum of electromagnetic radiation lying between the frequency limits of 3 kilohertz (kHz) and 300 gigahertz (GHz). Radio frequencies are grouped into bands. Radio signals travel through space in the form of waves. These waves vary in length, and each wavelength is associated with a particular radio frequency. Agencies are assigned specific frequencies by the FCC.

[7] Public Safety Wireless Advisory Committee (PSWAC), Final Report of the Public Safety Wireless Advisory Committee to the Federal Communications Commission and the National Telecommunications and Information Administration, September 11, 1996, https://www.ntia.doc.gov/legacy/osmhome/pubsafe/PSWAC_AL.pdf.

transition; as a result, the spectrum was not immediately available to public safety.

During the terrorist attacks of 9/11, public safety agencies could not communicate with each other. Radio channels were overloaded, and the multiple public safety agencies that responded to the events were operating on various frequencies and could not interoperate. In reviewing the events, the 9/11 Commission confirmed that New York City police and firefighters, operating on different frequencies, and using different channels, could not communicate, which hindered response. Further, the 9/11 Commission noted that systems and channels were overloaded by the large number of responders trying to communicate that day. The Commission called for the expedited and increased assignment of radio spectrum for public safety use.[8]

In the act, Congress provided a large swathe of spectrum, in a single band to enable agencies to communicate and coordinate during emergencies. In the act, Congress reallocated the 700 MHz D Block spectrum from commercial to public safety use.[9] This 10 MHz block was adjacent to an existing block of public safety spectrum that, when combined, provided 20 MHz of prime broadband spectrum to public safety use.

This combined spectrum (now called FirstNet spectrum or Band Class 14 spectrum) is serving as the foundation for the nationwide public safety broadband network created under the act. Figure 1 provides an overview of the FirstNet spectrum.

This FirstNet (Band Class 14) spectrum is to be used to build out a dedicated public safety broadband network. Broadband is commonly used to mean any high-speed access to the internet.[10] A broadband network for

---

[8] The 9/11 Commission, *The 9/11 Commission Report*, July 22, 2004, p. 397, https://www.9-11commission.gov/report/ 911Report.pdf.

[9] The 700 MHz spectrum is spectrum available for both commercial and public safety use. The 700 MHz spectrum is considered prime spectrum. Its propagation characteristics allow the 700 MHz signals to penetrate buildings and walls easily and to cover larger geographic areas with less infrastructure. Commercial providers can use this spectrum for mobile broadband services for smartphones, laptops, and tablets.

[10] The FCC defines broadband as a connection with speeds of 25 megabits down and 3 megabits up.

public safety is to provide public safety users with high-speed access to data (e.g., video transfer, internet access) and specialized features and applications (e.g., prioritization of users, location-based services, device management), that are not available on current commercial networks or public safety radios.[11]

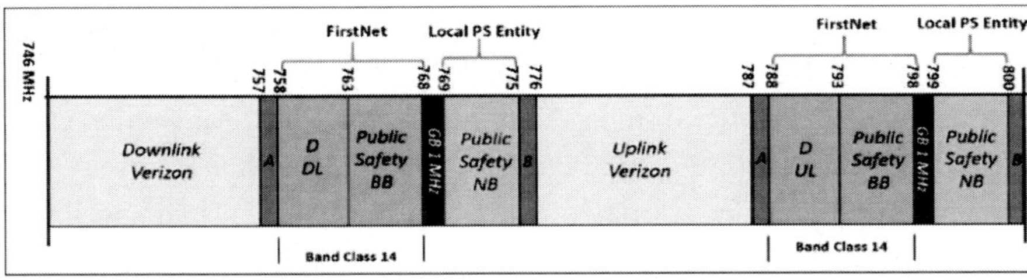

Source: Kumbhar and Guvenc, *A Comparative Study of LMR and LTE-Based Public Safety Communications,* 2015, Florida International University, Miami Florida, p. 2, researchgate.net.

Notes: Radio spectrum is often grouped in pairs; a block of spectrum in a lower frequency is paired with a block of spectrum in an upper frequency band. One band transmits in one direction (e.g., end-user to the network); this is known as the uplink (UL). Its corresponding pair transmits in the opposite direction (e.g., network to end-user); this is known as the downlink (DL). The green blocks show the paired spectrum that will be used for FirstNet. Also shown is the existing public safety broadband (BB) spectrum (763-768, 793-798 MHz) that was previously allocated for public safety use, and the D Block (758-763, 788-793 MHz) that was allocated to public safety under P.L. 112-96. One pair provides 10 MHz of spectrum (758-768 MHz) to serve as the downlink (DL) for the FirstNet network. The other provides 10 MHz (788-798 MHz) to serve as the uplink. The 10 MHz for the uplink and the 10 MHz for the downlink provide 20 MHz for the FirstNet network. As stipulated in the act, FirstNet holds the license to this paired spectrum (in green). The yellow blocks show narrowband (NB) spectrum assigned to public safety, used for voice communications and nationwide interoperable communications. Guard Bands (GB) of 1 MHz are placed between broadband, narrowband, and commercial carrier spectrum to prevent interference.

Figure 1. D Block and Public Safety Broadband Spectrum.

[11] Public safety radios provide access to mission critical voice features, such as push-to-talk features, group calling over the radio, direct mode communications which allow responders to communicate off-network with others who are in their vicinity—a useful feature when responders are out of range of their network or when systems are down.

While many public safety agencies already use commercial broadband services (e.g., cell phones, laptops), they access these services through commercial providers and commercial networks. During times of heavy use or congestion, which often occurs during emergencies, public safety users face the same constraints as commercial customers—loss of service, network unavailable, inability to make a call due to the high volume of traffic. Public safety users require higher levels of security, reliability, and redundancy to effectively perform their emergency response and life-saving functions. LMR systems have been built to serve public safety users; these are public safety-only systems that provide users with immediate access to the agency's communication system and specialized features for public safety users. Most have, over time, been strengthened and reinforced to ensure the communication system is reliable, resilient, secure, and protected from natural and man-made disasters. Commercial networks are built to different standards.

Therefore, at any given time, public safety agencies could be using:

- LMR (radio) systems that are owned and operated by individual public safety agencies, serve a specific agency or set of agencies within a certain geographic area, operate on various frequencies across the spectrum, and may or may not be interoperable with other LMR systems in the area. These systems offer reliable *voice* capabilities, are tried and tested in emergencies, have been hardened (i.e., reinforced) to withstand emergencies, and often include back-up capabilities if the network goes down;
- Commercial broadband/cellular networks that are owned and operated by commercial providers, offer nationwide coverage, enable voice communications between users, and offer high-speed access to data and applications, but do not always provide the levels of accessibility, reliability, security, and features that public safety users need; and
- FirstNet is to provide the benefits of commercial broadband networks (e.g., priority access to high-speed data, location-based services, incident management tools), with added reliability,

security, and redundancies that public safety needs. At first launch, however, FirstNet will not be able to offer the mission critical voice capabilities that are available through LMR systems; industry standards were approved in May 2016, and device development and testing is underway.[12] Therefore, LMR systems will need to be maintained to ensure public safety has access to mission critical voice capabilities.

State and local public safety agencies leverage different systems to communicate during incidents. FirstNet is intended to supplement these systems, to provide public safety users with dedicated spectrum, added broadband capabilities, and advanced technologies to increase situational awareness and enhance response.

The Value of the FirstNet Spectrum

Congress allocated 20 MHz of valuable broadband spectrum to public safety. As stipulated in the act, FirstNet holds the license to the 20 MHz of spectrum. The act authorizes FirstNet to enter into a public-private partnership and a covered leasing agreement—an agreement to construct, manage, and operate the network. The act also permitted access to network capacity on a secondary basis (i.e., access to the spectrum when it is not in use by public safety).[13]

In its RFP, FirstNet made all 20 MHz of the FirstNet (Band Class 14) spectrum available to the winning bidder. The awardee is permitted to monetize (i.e., earn revenue from) all 20 MHz of the spectrum. As the

---

[12] *Mission critical voice* capabilities—specialized features that first responders have on their public safety radios and use during emergencies (e.g., push-to-talk features, ability to communicate off-network, emergency calling for first responders)—are not yet available on FirstNet devices. Industry standards for mission critical push-to-talk over LTE (broadband) networks were approved in May 2016 by 3GPP; device development and testing are underway. Therefore, at first launch, mission critical voice features will not be available on the FirstNet network. Public safety agencies will need to maintain their current LMR systems to ensure first responders have access to mission critical voice features.

[13] Sec. 6208 (a) (2)(B)(i) of P.L. 112-96.

winning bidder, AT&T can generate revenue from public safety users which must be used to support and improve the network. AT&T can also generate revenue from other users (e.g., secondary users, commercial customers) when the spectrum is not in use by public safety. Some experts estimated that public safety agencies will use approximately 1% of the FirstNet spectrum, allowing AT&T to use a large portion of the spectrum for commercial use.[14]

In winning the contract, AT&T gained access to 20 MHz of unencumbered nationwide spectrum.[15] Further, the spectrum is not limited to a defined geographical area.[16] Additionally, the spectrum is licensed for *public safety* use, which permits the use of higher powered devices that can provide wider coverage. And, the spectrum does not count against spectrum holding limits set for wireless carriers by the FCC, allowing AT&T to bid on other spectrum.[17] These features add to the value of the FirstNet spectrum, which some experts have estimated to be $8 billion.[18]

AT&T noted that the award of FirstNet aligns with its plan to deploy a commercial 5G network, and that it sees efficiencies that can be achieved from "climbing the tower once" to install equipment for FirstNet, and for its own 5G network;[19] further, the award of FirstNet has provided AT&T with a wider fiber footprint (i.e., a nationwide footprint that includes access

---

[14] Trefis Team, "How Much Does AT&T Stand to Gain from FirstNet?," *Forbes*, December 12, 2017, https:// http://www.forbes.com/sites/greatspeculations/2017/12/12/how-much-does-att-stand-to-gain-from-firstnet/ #78dae0772997.

[15] Unencumbered typically means that the spectrum has been cleared; there are no licensees operating on the spectrum.

[16] The FCC issues licenses to state and local public safety agencies that define not only the frequency on which the agency can operate, but also the geographical area. FirstNet spectrum is not limited to a certain geographic area.

[17] Department of the Interior, *FirstNet Nationwide Public Safety Broadband Network (RFP), Section C: Statement of Objectives*, Solicitation Number: D15PS00295E, Herndon, VA, January 8, 2016, p. C-2.

[18] See Seamus Conwell, "This $15 Billion Spectrum Contract Could Be an 'Issue' for the AT&T Time Warner Deal," CNBC, March 3, 2017, https://www.cnbc.com/2017/ 03/03/this-15-billion-spectrum-contract-could-be-an-issue-for-the-att-time-warner-deal.html. The text says that analysts, including analysts at UBS and Wells Fargo, have estimated the value of the FirstNet spectrum to be $8 billion.

[19] 5G is the next generation wireless technology that promises to provide faster connections, greater bandwidth, extra capacity, quicker download and upload speeds, etc. 5G networks may power self-driving cars, Smart Cities, and connect sensors and other machines. 5G technologies involve the deployment of thousands of small cells (miniature cell towers) on local infrastructure, such as lamp posts, buildings, and other structures.

to fiber that can connect infrastructure, networks, and devices), which AT&T believes will serve as an advantage during 5G network deployment.[20]

Network Architecture

The act required FirstNet to ensure the establishment of a nationwide, interoperable public safety broadband network (see Figure 2), based on a single, nationwide architecture that evolves with technological advancements. Under the act, the network consists of two components:

1. A core network that:
   - consists of national and regional data centers, and other elements and functions that may be distributed geographically, and based on commercial standards; and
   - provides the connectivity between (i) the radio access network, and (ii) the public Internet or the public switched network, or both.
2. A radio access network (RAN) that:
   - consists of all cell site equipment, antennas, and backhaul equipment, based on commercial standards, that are required to enable wireless communications with devices using the public safety broadband spectrum;
   - shall be developed, constructed, managed, maintained, and operated, taking into account the plans developed under the state and local implementation grant program.[21]

---

[20] Securities and Exchange Commission, "Partial Transcript of Q1 2017 AT&T Inc. Earnings Call," April 25, 2017, https://www.sec.gov/Archives/edgar/data/732717/000073271717000049/r425.htm.

[21] State and Local Implementation Grant Program: https://www.ntia.doc.gov/sligp/program_information.

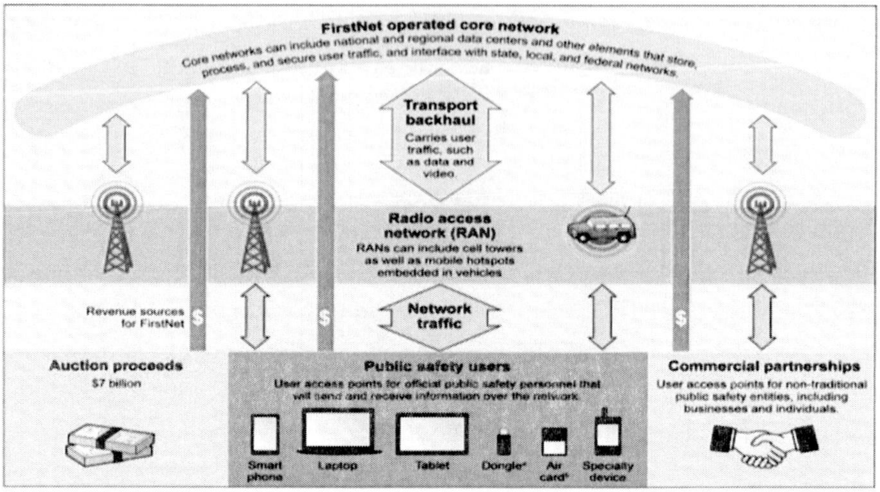

Source: U.S. Government Accountability Office, Public-Safety Broadband Network: FirstNet Should Strengthen Internal Controls and Evaluate Lessons Learned, GAO-15-407, April 28, 2015.

Note: Figure 2 depicts the elements of the FirstNet network, and the flow of information between the elements. For example, Public safety users can use a variety of devices to access the FirstNet network. The message flows from the public safety user's device to the radio access network (RAN), which include cell towers, mobile units in cars, and backhaul (e.g., fiber that can transport messages). The messages are carried to the core network which can include regional data centers and other infrastructure that processes, stores, and secures the data, and can enable interconnections with federal, state, and local networks, as well as the internet. Figure 2 also shows revenue that will be used to support the network including auction proceeds ($7 billion), and the private partner contribution, which will be $40 billion over the 25-year span of the contract.

Figure 2. Elements of the FirstNet Network.

FirstNet, with its private partner (AT&T), is responsible for building the core network. The core is a key component for ensuring that users have a consistent experience nationwide, and that the network is interoperable. Through the core, FirstNet can ensure a centralized approach to managing access, uniform user profiles, applications and services; further, FirstNet can consistently manage information and users and ensure the security of network. The RAN portion of the network (i.e., cell towers, hot spots embedded in vehicles) is to be deployed state-by-state. Traffic is to flow from devices (e.g., smart phones, tablets) through the RAN and is to backhaul to the core network over satellite or other wireless infrastructure.[22]

The act requires the network to comply with minimum technical interoperability requirements, set by an Interoperability Board, and based on commercial standards for Long Term Evolution (LTE) service.[23] The act encourages the use of commercial and government-owned infra-structure to speed the deployment of the network, and permits FirstNet to enter into roaming agreements, which would allow users to make and receive calls when traveling outside the geographic coverage area of the FirstNet network. The act also requires rural coverage milestones be set throughout each phase of deployment, to ensure agencies in rural areas have access to FirstNet.

Major Developments in 2017

On March 30, 2017, FirstNet awarded AT&T with a 25-year contract to build, operate, and maintain the nationwide public safety broadband network.[24] Under the agreement, FirstNet will provide $6.5 billion to AT&T over the first five years, and access to 20 MHz of broadband spectrum (also called FirstNet spectrum or Band 14 spectrum, discussed above). AT&T will provide FirstNet with access to its existing infrastructure, valued at more than $180 billion, and with $40 billion over the life of the contract to support the public safety network.

Under the contract, AT&T will build a 4G Long Term Evolution (LTE) network for public safety users. AT&T has a five-year deployment plan, which includes the development and deployment of the core network,

---

[22] Backhaul refers to links that carry user traffic (e.g., voice, data, video), and signaling from base stations to the core.

[23] LTE is a standard for wireless communications set by 3GPP that indicates a streamlined network architecture, roaming on other networks, and faster connections for users accessing applications and multimedia on mobile devices.

[24] The award was delayed for several months, after Rivada Mercury unsuccessfully challenged the selection process in the U.S. Court of Federal Claims. See *Rivada Mercury, LLC v. The United States of America and AT&T Corp.*, 16- 1559C (U.S. Court of Federal Claims 2017), https://ecf.cofc.uscourts.gov/cgi-bin/show_public_doc?2016cv1559-87-0.

and the deployment of the radio access network (RAN) in each state and territory. During the deployment period, AT&T has offered public safety users access to its commercial network with priority and preemption services. Priority access means that public safety users can access AT&T's commercial network during emergencies, when there is typically heavy use and congestion.

In June 2017, FirstNet/AT&T provided states with preliminary state plans, detailing how FirstNet/AT&T would deploy the Radio Access Network (RAN) portion of the network in each state (e.g., cell towers, mobile hot spots). The deployment plans were based on state and local needs, and other information collected during consultations between FirstNet and states over the past three years. Plans included information on the network architecture, including coverage and pricing— two critical elements driving state decisions to opt in or out of the network.[25] States provided feedback to FirstNet/AT&T, which was used to prepare the final state plan.

On September 29, 2017, FirstNet/AT&T released final plans to the states. As required in the act, FirstNet sent an official notice to each governor, which triggered a 90-day review of the state plan. Governors could accept the FirstNet/AT&T plan for the build-out of the network in their state (i.e., opt in), *or* issue a Request for Proposal (RFP) to build its own RAN (i.e., opt out). After consideration of the costs and benefits, all 50 states and 6 territories opted in.[26]

Additional developments in 2017 included:

- *Spectrum Relocation***.** The National Telecommunications and Information Administration (NTIA) worked with public safety agencies that were operating in the FirstNet spectrum (Band Class 14) to relocate those licensees to other 700 MHz spectrum, to clear

---

[25] Testimony of Mr. Curtis Brown, in U.S. Congress, Senate Committee on Commerce, Science, and Transportation, *An Update on FirstNet,* hearings, 115th Cong., 1st sess., July 20, 2017.

[26] It is important to note that while the governor or territorial leader may have opted in to FirstNet (i.e., allowed AT&T to deploy the network in the state or territory), state and local public safety agencies (e.g., state police, local fire departments) are not required to subscribe to the network (i.e., use the network).

the spectrum for the FirstNet network. NTIA awarded nearly $27 million through its Spectrum Relocation Grant Program to fund relocation costs. As of April 2018, 8 of 10 funded projects are complete.

- *Public Safety Broadband Research.* The National Institute of Standards and Technology (NIST) awarded $38.5 million under the Public Safety Innovation Accelerator Program, to 33 entities to conduct public safety broadband research. This program received $300 million funded from revenue generated by spectrum auctions authorized in the act. The $38.5 million represents the first round of public safety broadband research awards.
- *Next Generation (NG911) Grant.* In September 2017, NTIA and the Department of Transportation's National Highway Traffic Safety Administration (NHTSA) proposed rules for the NG911 program funded from revenue generated by spectrum auctions authorized in the act. The grant is aimed at improving state and local 911 technologies and services. NTIA and NHTSA are finalizing the rules and expect to release the grant notice in 2018.
- *State and Local Implementation Grant Program (SLIGP).* In 2017, NTIA announced a second round of funding (SLIGP 2.0). SLIGP 2.0 can fund personnel to help coordinate FirstNet activities, including identifying users, establishing governance structures, conducting data collection, and developing policies to increase data sharing across public safety systems. Every state and territory is eligible for SLIGP 2.0 funding. The grants have a 20% matching requirement, a two-year period of performance, and are to be awarded in 2018.
- *Tribal Consultation.* In October 2017, FirstNet released a *Tribal Consultation Policy*, establishing a nation-to-nation relationship between tribes and FirstNet, to ensure effective consultation with tribes during the build-out of the network. The policy was

developed in accordance with Executive Order 13175 which sets core principles for engaging with federally-recognized tribes.[27]

Issues for Congress

In 2017, Congress conducted oversight hearings on FirstNet.[28] Some Members questioned FirstNet, AT&T, and state representatives on state plans, the opt-in/opt-out process, factors affecting opt-in/opt-out decisions, and critical aspects of the FirstNet/AT&T network, including coverage (rural and non-rural); the security, reliability and redundancy of the FirstNet network; features available on the network; site hardening (e.g., strengthening and protecting critical infrastructure from disasters); and cost. These issues were also raised by states during the review of their state plans, and may indicate issues affecting public safety agencies' decisions to use the FirstNet network, which is essential to its long-term sustainability.

The act requires that the network be self-sustaining through user fees, but it does not require public safety agencies to subscribe to the network. Congress may provide oversight of FirstNet to ensure that the network is being deployed as required in the act and as intended by Congress. Congress may also monitor the deployment to ensure that FirstNet/AT&T are meeting public safety needs and providing features and services that will attract users to the network. Users are needed to ensure that the network will be self-sustaining, as required in the act, and to achieve the intent of the law—to provide a single, interoperable network that public safety agencies can use to communicate and coordinate during disasters.

In providing oversight, Congress may consider the following issues.

---

[27] Executive Order 13175, "Consultation and Coordination with Indian Tribal Governments," (Washington: GPO, 2000), https://www.gpo.gov/fdsys/pkg/FR-2000-11-09/pdf/00-29003.pdf.

[28] U.S. Congress, Senate Committee on Commerce, Science, and Transportation, Communications, Technology, Innovation, and the Internet, *An Update on FirstNet*, 115th Cong., 1st sess., July 20, 2017; and U.S. Congress, Senate Committee on Commerce, Science, and Transportation, Communications, Technology, Innovation, and the Internet, *Oversight of FirstNet: State Perspectives*, 115th Cong., 1st sess., November 1, 2017.

Transparency of AT&T Contract

The FirstNet RFP was released in January 2016. The RFP did not dictate how the network should be deployed; instead, the RFP asked bidders to provide their own innovative solutions to achieving sixteen high-level objectives included in the RFP by FirstNet.[29] AT&T was awarded the contract in March 2017 based on its innovative approach, technical capability, and financial stability; however, the contract was deemed proprietary and was not available for public view. Since there is no specific approach delineated in the RFP, and since the actual approach is not available for public review, it is likely difficult for Congress to understand how the FirstNet network will be deployed. A number of Members of Congress have sought clarity on several aspects of the network. For example, during recent congressional hearings, some Members asked how and where the FirstNet network will be deployed; how FirstNet/AT&T will meet rural coverage requirements in the act; and if the AT&T solution provides the security, reliability, and resiliency that public safety agencies need during emergencies.[30] Without access to the contract, or greater details on the deployment, it is likely difficult for Congress to ensure that the requirements in the act are being fulfilled, and that the intent of the law is met.

Further, the RFP required FirstNet/AT&T to develop state plans which detailed the deployment of the radio access network (RAN) in each state. The state plans provide information on the coverage provided by AT&T, the deployment schedule, the services and devices available to users, and the cost to users. FirstNet/AT&T released preliminary plans in June 2017. Like the contract, the state plans were also deemed proprietary. Plans were made available to select state officials through a secure portal that required

---

[29] For example, FirstNet required that the network serve users nationwide and be interoperable; be deployed quickly; be secure; and that a variety of devices must be offered, etc. The RFP did not provide "one right way" to achieve these objectives but allowed the bidders to present their own solution to achieving these objectives.

[30] U.S. Congress, Senate Committee on Commerce, Science, and Transportation, Subcommittee on Communications, Technology, Innovation, and the Internet, *An Update on FirstNet*, 115th Cong., 1st sess., July 20, 2017, https://www.commerce.senate.gov/public/index.cfm/hearings?ID=28BA6F3B-F540-4373-8DC8-7C42006FBC81.

users to accept lengthy terms and conditions, including a non-disclosure agreement. Since the state plans were not released, it is not possible to analyze the plans, or to compare coverage, services, and pricing between states. Without insight into state plans, it will likely be difficult for Members of Congress to understand the FirstNet deployment plan for their state, to analyze how funding and other resources have has been distributed across states, or to ensure that requirements in the act are met.

This $6.5 billion, technically and financially complex project was proposed by Congress in response to the communication challenges experienced during 9/11 and other incidents. Given its size and the importance of the network to the life and safety of first responders and citizens, the current proprietary nature of the detailed plans for the project may hinder the ability of Congress to conduct effective oversight. Congress may consider requiring FirstNet to release the contract and state plans, or to provide greater detail on the network, so that Congress can ensure that the FirstNet/AT&T approach meets the requirements in the act, and the intent of the law: to provide a nationwide network for public safety to communicate and coordinate during response.

Core-to-Core Interoperability

Throughout the development of the FirstNet RFP, public safety advocates and industry experts debated whether there should be one nationwide vendor for FirstNet, or if there could be multiple vendors, and multiple interconnected cores that comprise the FirstNet network.[31] After receiving extensive input from public safety advocates and industry experts, FirstNet/AT&T decided on a comprehensive solution (i.e., one-vendor, nationwide solution), which it included in its RFP. After the award of the RFP, several commercial providers advocated for the

[31] CRS Report R42543, *The First Responder Network (FirstNet) and Next-Generation Communications for Public Safety: Issues for Congress*, by Lennard G. Kruger.

interconnection of other cores to the FirstNet/AT&T core.[32] This would allow commercial providers to retain their public safety customer base and allow users access the FirstNet network, when needed.

AT&T argued that the RFP required the "provisioning of a [single] nationwide core," and did not allow for multiple core networks, and integrating multiple cores into the nationwide network would increase cyber risks.[33] FirstNet agreed with AT&T and re-stated the need for a single core, "to reduce the risk of complications inherent in a multi-core architecture (operated by distinct entities), such as operational complexity, security complexity, and increased latency."[34]

While the idea of core-to-core interoperability seemed to be rejected, the issue resurfaced when Verizon announced in August 2017 that it would be building its own dedicated, public safety core network for public safety customers.[35] Like AT&T's core, the Verizon core plans to operate separately from its commercial core, and provide first responders with access to the company's extended LTE network, and priority and preemption services. Verizon intends to also make available multi-band devices that will provide access to any Band Class 14 RAN deployed by FirstNet and committed (as did AT&T) to investing in new broadband technologies for its public safety customers.[36] Verizon appears to be

---

[32] See Letter from Mr. Declan Ganley and Mr. Joseph J. Euteneuer, Rivada Networks LLC, to Ms. Marlene Dortch, Secretary, FCC, June 12, 2017, https://ecfsapi.fcc.gov/file/106122435010907/ FCC%20ex%20parte% 20letter%20 FINAL%206-12-17.pdf.; and see Verizon blog, "Verizon to Build Dedicated Network Core for Public Safety," August 16, 2017, http://www.verizon.com/about/news/verizon-build-dedicated-network-core-public-safety.

[33] U.S. Congress, House Committee on Energy and Commerce, Subcommittee on Communications, Technology, Innovation, and the Internet, *Oversight of FirstNet: State Perspectives*, 115th Cong., 1st sess., November 1, 2017, https://energycommerce.house.gov/hearings/oversight-firstnet-state-perspectives/.

[34] FirstNet, "Facts About FirstNet: Our Nation's Public Safety Broadband Network," https://www.firstnet.gov/facts.

[35] Letter from Donald Brittingham, Verizon, to Subcommittee on Communications and Technology, October 31, 2017, http://docs.house.gov/meetings/IF/IF16/20171101/106569/HHRG-115-IF16-20171101-SD007-U7.pdf.

[36] See Verizon advertisement during 2018 Super Bowl, https://www.youtube.com/watch?v=ipx4Wu5P_lE.

offering a parallel service to FirstNet and continues to advocate for interconnections with the FirstNet/AT&T core.[37]

FirstNet has stated that its one-vendor approach will allow it to achieve efficiencies in management, procurement, deployment, and operation of the network. Some public safety advocates supported this approach;[38] others have argued that competition is good—that it will reduce costs, increase options, and spur innovation for public safety users.[39]

FirstNet and AT&T have expressed concern that the integration of other networks into the FirstNet network may affect security, interoperability, and sustainability of the network.

- *Security:* FirstNet and AT&T have stated that integrating other core networks that may not adopt and enforce the same technical requirements, protocols, training, and best practices could jeopardize security.[40]
- *Interoperability:* Verizon argued that the Technical Advisory Board created under the act to advise FirstNet on interoperability, *allowed* interoperability through "core-to-core interconnection and mutual automatic roaming arrangements...."[41] Further, Verizon stated that secure interoperability is feasible, and that a competitive marketplace spurs innovation.[42] AT&T noted interoperability between cores may be *technically* feasible, but it has not yet been

---

[37] Letter from William H. Johnson, Verizon, to Ms. Marlene Dortch, Secretary, FCC, July 24, 2017, https://ecfsapi.fcc.gov/file/10724307201497/2017%2007%2024%20Verizon%20 FirstNet%20FCC%20letter.pdf.

[38] APCO, "APCO Welcomes Publication of FirstNet's RFP," January 13, 2016, https://psc.apcointl.org/2016/01/13/ apco-welcomes-publication-of-firstnets-rfp/.

[39] Testimony of Mr. Robert LeGrande, in U.S. Congress, House Committee on Energy and Commerce, Subcommittee on Communications and Technology, *Oversight of FirstNet: State Perspectives,* hearings, 115th Cong., 1st sess., Nov. 1, 2017, http://docs.house.gov/ meetings/IF/IF16/20171101/106569/HHRG-115-IF16-20171101-SD001-U1.pdf.

[40] FirstNet, https://firstnet.gov/content/firstnet-will-provide-public-safety-users-priority-access-network#Security.

[41] Letter from William H. Johnson, Senior Vice President, Federal Regulatory and Legal Affairs, Verizon, to Ms. Marlene Dortch, Secretary, FCC, July 24, 2017, https://ecfsapi.fcc.gov/ file/10724307201497/2017%2007%2024%20 Verizon%20FirstNet%20FCC%20letter.pdf.

[42] Mike Maiorana, Senior Vice President, Verizon Enterprise Solutions, "Interoperability Will Work for Public Safety," *LinkedIn*, September 28, 2017, https://www.linkedin.com/ pulse/interoperability-work-public-safety-michael-maiorana/.

attempted between commercial carriers, and brings security risks; further, AT&T argued that this approach does not serve the requirement in the act to create an interoperable nationwide network based on a single, nationwide architecture.[43] FirstNet agreed with AT&T, noting that there is uncertainty in how core-to-core connections would affect interoperability and security.[44]

- *Sustainability:* In its 2017 Report, GAO noted that if commercial carriers compete, it may affect the state and local adoption of (and subscription to) the network, which could affect the long-term sustainment and improvement of the network. Verizon has already amended its service offerings to include priority service and a preemption offering to public safety users, in order to sustain its customer base. While the interoperability of cores may spur competition and reduce costs, it may also negatively affect subscriptions to FirstNet and the long-term viability of the network.

In the act, Congress required the network to be secure, interoperable, and self-sustaining. FirstNet/AT&T has adopted policies to ensure those requirements are met. However, questions on core-to-core interoperability remain. The development of a separate core by Verizon—which has stated that it services two-thirds of the public safety market—may affect subscribership to the FirstNet network, and threaten its long-term sustainability. Congress could mandate integration of the cores to increase the number of users on the network; however, it may seek to clarify if AT&T is planning on integration other cores, and to gain more information on the financial and security impact of integrating other cores into the

---

[43] Donny Jackson, "AT&T Exec Discusses Core-to-Core Interoperability, Verizon Proposal, FirstNet Differentiators," *Urgent Communications*, August 17, 2017, http://urgentcomm.com/public-safety-broadbandfirstnet/att-exec-discusses-core-core-interoperability-verizon-proposal-first.

[44] Donny Jackson, "Verizon Meets with FirstNet, but Interoperability Questions Remain," *Urgent Communications*, October 3, 2017, http://urgentcomm.com/ntiafirstnet/verizon-meets-first net-interoperability-questions-remain.

FirstNet core.[45] Congress may also encourage and support subscribership to the FirstNet network through other means; for example, Congress could provide grant funding to state and local public safety agencies to purchase FirstNet devices.

Services Beyond the Initial Five-Year Deployment

By January 2018, all 50 states and 6 territories accepted the FirstNet/AT&T plan to deploy the network in their state, although some opted in reluctantly.[46] Many states opted in and noted that there were outstanding items that were not fully addressed in state plans (e.g., future coverage, site hardening, improvements), that needed to be discussed with FirstNet and AT&T and addressed either in the five-year initial deployment or in the out-years. FirstNet and AT&T stated that they could not address all state and local public safety needs during the initial five-year deployment, but that they would continue to work with states to address outstanding issues.

Congress may continue its oversight of the FirstNet network through the initial five-year deployment of the network, and after. FirstNet has a contract with AT&T for 25 years. The act requires that revenues earned from public safety users be invested back into the network for improvements. Congress may clarify with FirstNet and AT&T how and where revenues will be used to improve the network. Congress may consider inviting state and local public safety agencies to testify at oversight hearings to hear whether issues not fully addressed in state plans or addressed in the initial five-year deployment (e.g., coverage, rural

[45] The act requires partnerships with rural providers which may own and operate their own cores; several pilot projects were funded in 2012 to test LTE technologies in anticipation of FirstNet—some of these projects deployed their own cores which may need to be interconnected to FirstNet's core; additionally, there may be unique telecommunications agreements (e.g., providers that service the Pacific territories) that may require core-to-core interconnections.

[46] Letter from Mark S. Ghirlarducci, Director, CalOES, to Michael Poth, CEO, FirstNet, December 28, 2017, http://www.oesnews.com/wp-content/uploads/2017/12/California_FirstNet_Decision_Letters_2017-12-28.pdf.

coverage, rural partnering agreements, site hardening, resiliency, security), are being adequately addressed in the out-years.

Coverage

Coverage has been a key issue for public safety since the network's inception. Without coverage, public safety agencies will not be able to use the FirstNet network to coordinate and communicate with others during emergencies. And, without coverage, agencies will not subscribe to the FirstNet network, hindering FirstNet's ability to ensure the network is self-sustaining as required under the law. In the act, there are no quantitative coverage requirements; however, the act does require substantial rural coverage milestones be set with each phase of deployment, and requires FirstNet to consult with state, tribal, and local entities regarding coverage needs.

In the RFP, FirstNet included two major provisions on coverage—a short term requirement to provide public safety with access to nationwide coverage within six months of award, to speed adoption of broadband, as encouraged under the act. And, a requirement to provide a longer-term, five-year deployment plan, with rural coverage milestones for each phase of deployment, as required under the act. While the provisions in the RFP reflect the requirements in the act, the proprietary nature of the contract and the state deployment plans makes it difficult to determine if the requirements of the law are being met, how the resources provided (e.g., $6.5 billion and 20 MHz spectrum) are being used, and if the network is being deployed as intended by Congress.

For the short-term deployment plan, FirstNet required the contractor to provide nationwide coverage, using the Band Class 14 spectrum (FirstNet spectrum) or other spectrum (non-Band Class 14), within six months of award. This provision allows the contractor to provide nationwide coverage by either building out the new nationwide network on Band Class 14 spectrum (FirstNet spectrum) or to provide nationwide coverage using their own spectrum and networks. This provision was intended to speed delivery of broadband to public safety agencies, as encouraged under the act, and to offer a nationwide solution while the FirstNet network was being built, which is expected to take five years for the initial deployment.

After the contract was awarded, AT&T announced that it would open *all* its spectrum bands that are LTE-enabled (i.e., all of its broadband spectrum) for public safety use and provide priority and preemption services to public safety users who subscribed to the FirstNet network. This would provide public safety users with immediate access to broadband services, access to a nationwide network, and assurances that public safety communications would be prioritized on AT&T's commercial network during emergencies.

Some proponents viewed the announcement as an opportunity to access commercial broadband services sooner, before the deployment of the FirstNet network, with the assurances that public safety communications would be prioritized on the AT&T network. Others saw AT&T's offer as a means to increase capacity (e.g., enable more users on the network), and to relieve congestion on crowded radio channels. Some argued that AT&T's approach could provide public safety agencies with flexibility to leverage new technologies (e.g., 5G) that may operate outside the 700 MHz band, as they become available.[47] Critics asserted that AT&T is using its commercial network as a base for the nationwide public safety broadband network—"rebranding" the AT&T network as the FirstNet network,[48] instead of building a dedicated public safety broadband network in Band Class 14, as intended in the act.[49]

AT&T stated that it is building out Band 14 as agreed to in the contract; this statement is supported by the RFP which requires that, six months from award, the contractor shall provide nationwide coverage on

---

[47] Donny Jackson, "Unanswered Questions Loom as Governors Prepare to Make 'Opt-In/Opt-Out' FirstNet Choices," *Urgent Communications*, October 4, 2017, http://urgentcomm.com/blog/unanswered-questions-loom-governors-prepare-make-opt-inopt-out-firstnet-choices.

[48] Albert J. Catalano, "Has FirstNet Rebranded AT&T's Network as the Nationwide Public Safety Broadband Network?," *National Law Review*, August 29, 2017, https://www.natlawreview.com/article/has-firstnet-rebranded-att-s-network-nationwide-public-safety-broadband-network.

[49] Donny Jackson, "Unanswered Questions Loom as Governors Prepare to Make 'Opt-In/Opt-Out' FirstNet Choices," *Urgent Communications*, October 4, 2017, http://urgentcomm.com/blog/unanswered-questions-loom-governors-prepare-make-opt-inopt-out-firstnet-choices.

Band Class 14 *or* Non-Band Class 14. However, the discussion around this topic raised questions as to when and where AT&T *would* be deploying Band Class 14 (FirstNet) network, and how the 20 MHz of spectrum allocated to public safety is being used.

In a July 2017 hearing of the Senate Subcommittee on Communications, Technology, Innovation, and the Internet, Senators asked where the Band Class 14 infrastructure would be deployed, and percentage of land covered.[50] AT&T stated that Band Class 14 will be deployed in areas where AT&T determines there is a need for additional capacity but declined to provide details on the deployment due to proprietary reasons. In later statements, AT&T later reported that the Band Class 14 network (i.e., the FirstNet network) will be deployed not only where there is a need for greater capacity (i.e., areas of high-density), but also on new sites to cover rural areas that are unserved or underserved.[51] AT&T also stated that deployable assets would be used to provide coverage to areas, when needed.

AT&T stated that they will cover more than 99% of the population, and 76.2% of the geography without deployable assets.[52] By combining the AT&T wireless LTE network with rural networks, deployable assets,[53] and additional satellite technology, FirstNet/AT&T stated that it will be able to cover more than 99% of the U.S. geography.[54]

---

[50] U.S. Congress, Senate Committee on Commerce, Science, and Transportation, Subcommittee on Communications, Technology, Innovation, and the Internet, *An Update on FirstNet*, 115th Cong., 1st sess., July 20, 2017, https://www.commerce.senate.gov/public/index.cfm/hearings?ID=28BA6F3B-F540-4373-8DC8-7C42006FBC81.

[51] Sandra Wendelken, "AT&T Exec: FirstNet Contract Requires Undisclosed Number of Sites to Be Built," *Mission Critical*, December 12, 2017. https://www.rrmediagroup.com/Features/FeaturesDetails/FID/808, https://www.rrmediagroup.com/Features/FeaturesDetails/FID/808.

[52] FirstNet, "Nationwide Coverage," https://www.firstnet.com/coverage.

[53] This may include cell-on-wheels (COW), cell-on-light-truck (COLT) or other deployable coverage options.

[54] U.S. Congress, Senate Committee on Commerce, Science, and Transportation, Subcommittee on Communications, Technology, Innovation, and the Internet, *An Update on FirstNet*, Response to Questions for the Record submitted by Chris Sambar to Sen. Thune, 115th Cong., 1st sess., July 20, 2017, https://www.commerce.senate.gov/public/_cache/files/3c2c1b6b-8e2d-4275-a0a8-d09f534effbc/7A2F902BCF0DE992179D2AA6AB6ECE48.mr.-sambar-07.20.17- senate-cst-qfrs.pdf.

While deployable solutions can offer coverage immediately following emergencies, they are not counted as persistent coverage toward rural milestones, nor do they serve as permanent communication solutions. Deployable assets take time to deploy and can be moved to serve another jurisdiction or agency if an emergency arises.

Without detailed information on the deployment, it is likely difficult for Congress to determine if FirstNet and AT&T are using the resources provided by Congress as intended, if the requirements in the law are being met, and if the network is being deployed as intended by Congress. Congress may request additional detail on FirstNet coverage, where the FirstNet network will be deployed and where it will not, and which regions will be served with persistent coverage versus deployable coverage.

Rural Coverage

The RFP required the inclusion of substantial rural coverage milestones with each phase of deployment, as required in the act. However, the contract allowed bidders to submit their own innovative solutions to meeting these requirements, and to set the base against which these milestones must be achieved. For example, within 12 months of award, the contract requires "Achievement of 20% of contractor's proposed Band 14 coverage in rural areas."[55] The contractor was permitted to set the base against which the milestones must be achieved. And, because the contract and the state deployment plans were deemed proprietary, the coverage offered by the contractor is not available for public view; as a result, there does not appear to be enough information to determine if the milestones are "substantial," as required under the act.

While the targeted milestones are a useful gauge for FirstNet to measure progress, the milestones are anchored to the contractor's proposed Band 14 coverage, which has been deemed proprietary and is not available for public view. Without knowing the nationwide Band 14 coverage

---

[55] Department of the Interior (on behalf of FirstNet), *FirstNet Nationwide Public Safety Broadband Network (NPSBN)*, Request for Proposal (RFP), Target Timeline, Herndon, VA, January 13, 2016, pp. Section J-8, https://www.fbo.gov/ utils/view?id=bc4a63534ebd 1fd64cd01ff070f25603.

offered by AT&T, it is difficult to assess the validity of contract milestones, as they are a percentage of an unknown factor. Further, due to the proprietary nature of the contract and the state plans, it is difficult to determine which areas are included in the milestones (i.e., which areas will be covered) and which areas are not.

Congress may request more information on FirstNet coverage, rural milestones, and areas that will be covered under FirstNet, to determine if the "substantial rural coverage" requirements in the law are being met.

Rural Partnering Agreements

In the act Congress directed that "[t]o the maximum extent economically desirable, such proposals shall include partnerships with existing commercial mobile providers to utilize cost- effective opportunities to speed deployment in rural areas." FirstNet included a 15% partnering requirement in the RFP, which AT&T has stated that it will likely exceed.

The Competitive Carriers Association (CCA)—which represents rural and regional carriers—expressed interest in partnering with FirstNet. In a November 2017 letter to Congress, CCA encouraged FirstNet to work collaboratively with all rural partners. CCA stated that rural and tribal telecommunication companies are often the only providers in rural regions, serve as the first line of defense in emergencies, and can provide reliable back-up during times of outages.

CCA noted that establishing partnerships would strengthen rural carriers and limiting partnerships could "delay and reduce services for public safety users, and result in AT&T using spectrum and funding provided by Congress for public safety use to eliminate commercial competitors."[56]

Due to the proprietary nature of the contract, information on the rural partnering agreements, which could indicate where rural coverage may be

[56] U.S. Congress, House Committee on Energy and Commerce, Subcommittee on Communications and Technology, *Oversight of FirstNet: State Perspectives*, Letter for the Record, submitted by the Competitive Carriers Association (CCA), 115th Cong., 1st sess., November 1, 2017, http://docs.house.gov/meetings/IF/IF16/20171101/106569/HHRG-115-IF16-20171101-SD004-U4.pdf.

provided, is not available to the public. Congress may ask FirstNet where rural partnering agreements exist to ensure the requirement in the law concerning rural partnerships is met, and encourage FirstNet and AT&T to work with other interested partners and to expand service to rural regions.

Public Safety Grade

In the act, Congress requires the establishment of technical requirements to ensure the network is reliable, secure, resilient, and can withstand damage and disasters (e.g., hardening requirements). In 2013, FirstNet stated that the public safety broadband network will be "public safety grade" which, in simple terms, relates to standards by which public safety systems are built—with added requirements and features, to ensure the network keeps working in the harshest conditions, during emergencies and after. For example, public safety sites may have additional or extended back-up power, multiple paths to backhaul network traffic, and extensive site hardening to ensure communications systems are sustained during emergencies (e.g., tornados, earthquakes).

In a congressional hearing in July 2017, Senators raised questions regarding the reliability of the AT&T network, and whether the network was considered "public safety grade." In response, AT&T noted that there is variability in the definition of "public safety grade," and that its network is built to meet federal, state, and local laws, standards, and requirements. AT&T stated that its network is reliable, and when and if it experiences outages, AT&T has a robust deployable solution (i.e., extensive portfolio of deployable assets, emergency operations center, and customer support) that can be used to restore the network, as demonstrated in the 2017 hurricane season.[57]

---

[57] U.S. Congress, Senate Committee on Commerce, Science, and Transportation, Subcommittee on Communications, Technology, Innovation, and the Internet, *An Update on FirstNet*, 115th Cong., 1st sess., July 20, 2017, https:// energycommerce.house.gov/hearings/ oversight-firstnet-state-perspectives/.

Although commercial systems meet all federal, state, and local siting, building and maintenance laws governing telecommunication systems, they do not always meet the higher-level standards for availability and reliability of public safety systems.[58] Congress may consider requesting more information on AT&T's network since it has been offered to public safety users while the FirstNet network is being built; Congress may seek to ensure the AT&T network being used by public safety is secure, reliable, resilient, and has redundancies built in to ensure that it will withstand major events (e.g., hurricanes, tornadoes), and that public safety communications are protected.

Further, Congress may consider requesting more information on how FirstNet and AT&T will ensure that the FirstNet network will be reliable and resilient, and able to withstand damage or disasters. The National Public Safety Telecommunications Council (NPSTC), in conjunction with public safety organizations and the standards development organizations, has released "Defining Public Safety Grade Systems and Facilities," which included input from multiple public safety organizations, best practices and requirements on network hardening and network reliability.[59] NPSTC acknowledged that it would be difficult and costly to implement all the best practices throughout a nationwide data system, on every site, but the report provided public safety network needs, and benchmarks for achieving a true "public safety grade" network.

Similarly, AT&T stated that, "We need a public-safety-grade network that extends from the handsets to the central office," however, given the resources, "[it's] not reasonable to think every tower will be at public-safety-grade level ... there needs to be some ranking ...we are working with the states ... and with the PSAPs (public-safety answering points)."[60]

---

[58] U.S. Congress, House Committee on Energy and Commerce, Subcommittee on Communications and Technology, *Oversight of FirstNet: State Perspectives*, 115th Cong., 1st sess., November 1, 2017.

[59] NPSTC, *Defining Public Safety Grade Systems and Facilities*, May 22, 2014, http://www.npstc.org/download.jsp?tableId=37&column=217&id=3066&file=Public_Safety_Grade_Report_140522.pdf.

[60] Sandra Wendelken, "AT&T Exec Backtracks on Public Safety Grade Comments," *Mission Critical*, September 7, 2017, https://www.rrmediagroup.com/News/NewsDetails/NewsID/15933.

FirstNet and AT&T have stated that they will continue to work with states to identify critical assets that need to be hardened. Congress may seek to confirm that FirstNet and AT&T are adhering to the technical requirements established for the network under the law. Further, Congress may inquire how FirstNet/AT&T decides which sites are hardened, and how it expects future revenues will be used for public safety grade improvements.

The security, reliability, and resiliency of the FirstNet network are important to not only ensuring the network will continue to operate during and after major disasters, but also to gaining the buy-in of public safety users. In its July 2017 report, GAO noted that public safety stakeholders it interviewed raised concerns about several aspects of the network, including resiliency; reliability; redundancy; cybersecurity; frameworks for user identity, credentialing of users, and access management; and prioritization of users on the network. States raised similar concerns during the review of state plans.[61] Congress may consider requesting more information from FirstNet and AT&T on the FirstNet network's public safety grade features and when they will be available to public safety users, as these issues may affect state and local public safety agencies' decisions to join the network, the ability of the project to be self-sustaining, and the success of the network. Congress may also consider inviting state and local public safety users to testify at hearings on FirstNet to understand issues affecting subscribership.

Mission Critical Push-to-Talk (MCPTT)

Most public safety agencies use land mobile radios (LMR) to communicate. Commercial vendors have developed specialized features to meet the specific needs of public safety users during day-to-day operations

---

[61] See California's comments, requesting additional information on site hardening, http://www.caloes.ca.gov/PublicSafetyCommunicationsSite/Documents/FirstNetinCaliforniaStatePlanReviewResults.pdf, p. 12.

and emergencies. Mission critical push-to-talk voice (MCPTT)[62] includes several features that have been standardized for LMR devices and systems, including:

- Direct Mode (i.e., allows device-to-device communication with nearby users off-network, when out of network range, or if the network is down);
- Push-to-Talk (PTT) (i.e., the responder pushes a button to talk, and releases the button to listen, which promotes ease of use);
- Group call (i.e., enables one-to-many communications);
- Full Duplex Voice (i.e., enables responders to communicate with parties outside their own PTT systems, including citizens with emergencies, and other agencies);
- Alerting (i.e., indicates a responder has encountered a life threatening emergency);
- Talker identification (i.e., similar to caller identification); and
- Audio Quality (i.e., high audio quality to enable the listener to better understand the speaker).

Public safety users want these features on the FirstNet (Band 14) network and devices. However, the standards for MCPTT features (e.g., direct mode, PTT, alerting) for LTE devices and networks (including FirstNet) were not yet developed when the law was signed.[63] 3GPP—the international standards development organization—prioritized the development of technical standards for public safety broadband and declared that establishing common technical standards for commercial

[62] NPSTC, *Mission Critical Voice Communications Requirements for Public Safety*, 2011 http://www.npstc.org/download.jsp?tableId=37&column=217&id=2055&file=Mission%20 Critical%20Voice%20Fu.

[63] Dan Verton, "FirstNet Facing Commercial Device Reality, but Market and Security Concerns Persist," *fedscoop*, October 9, 2014, https://www.fedscoop.com/firstnet-facing-commercial-device-reality-but-security-concerns-persist/.

cellular and public safety features will ensure interoperability between different vendors and lead to a competitive equipment market.[64]

The standards for MCPTT over LTE networks were recently approved by 3GPP, but still need to be tested and validated, which will take time. Therefore, at first launch, the FirstNet (Band 14) network will not offer MCPTT voice. Responders should be able to talk on FirstNet/AT&T devices, but the MCPTT features—critical features for first responders—will not be available on the FirstNet network. FirstNet has noted that the network will primarily be a high-speed *data* network. "The network is expected to initially transmit data, video, and other high-speed features, such as location information and streaming."[65] FirstNet has said that it is expecting to be able to offer MCPTT technology across the network by March 2019.

As a result, state and local public safety agencies may continue to use their LMR systems, until all features, including MCPTT voice, are available from FirstNet. The delay of MCPTT features may impact agencies' willingness to subscribe to the network, and affect FirstNet's ability to ensure the network is self-sustaining, as required under the law.

FirstNet has begun to offer devices that will be able to operate on the FirstNet network, with modified features for public safety (e.g., proprietary push-to-talk, ruggedized devices). However, questions have been raised about these devices and the device policy.

Some Members of the House Energy and Commerce Committee raised questions regarding whether FirstNet devices can successfully roam onto rural wireless networks. AT&T confirmed that responders will be able to roam across various networks because they all operate on 3GPP standards. However, it is not clear if FirstNet users can roam on *all* rural networks, or just those carriers that have agreements with AT&T.

Some carriers have raised questions on device interoperability. The Competitive Carriers Association (CCA) asserts that the AT&T devices are "compatible only with AT&T's boutique Band Class 17 ... and may not be

[64] See 3GPP webpage on Public Safety: http://www.3gpp.org/news-events/3gpp-news/1455-Public-Safety.

[65] See FirstNet Network webpage: https://firstnet.gov/network/lmr.

compatible with the public safety spectrum in Band Class 14, [and] ... unable to access a connection in areas outside of AT&T's commercial network." [66]

Public safety advocates have raised questions on whether other carriers that install Band 14 into their devices would be able to access FirstNet services.[67] It is not clear if users would have to pay for two subscriptions—one for the primary carrier, and another to FirstNet; or if non-AT&T device-holders would be able to roam onto all of AT&T bands or just Band 14, which could affect the ability of agencies to interoperate.

At a June 2017 hearing, witnesses testified on the number of responders that use their own personal devices for work, such as volunteer firefighters.[68] FirstNet/AT&T stated it would establish a Bring Your Own Device (BYOD) policy for those users, and that it would certify personal devices for compatibility, interoperability, and security.

FirstNet is working to expedite the development of LTE Band 14 devices, and has testified that it supports the development of standards-based, non-proprietary devices, equipment, and applications that will be interoperable across various networks. FirstNet is facilitating the testing of devices at its Innovation Lab in Boulder, CO, and it is playing an active role in standards-development, as required under the act. In 2017, FirstNet attended Mobile World Congress, where they arranged market research meetings with LTE device vendors to discuss the optimum integration of LTE Band 14 within their products and to promote key public safety

---

[66] U.S. Congress, House Committee on Energy and Commerce, Subcommittee on Communications and Technology, *Oversight of FirstNet: State Perspectives*, Letter for the Record, submitted by the Competitive Carriers Association (CCA), 115th Cong., 1st sess., November 1, 2017, http://docs.house.gov/meetings/IF/IF16/20171101/106569/HHRG-115-IF16-20171101-SD004-U4.pdf.

[67] Andrew Seybold, "Public Safety Advocate: FirstNet or SecondNet?" *All Things FirstNet*, August 17, 2017, http://allthingsfirstnet.com/public-safety-advocate-firstnet-or-secondnet/.

[68] U.S. Congress, Senate Committee on Commerce, Science, and Transportation, Subcommittee on Communications, Technology, Innovation, and the Internet, *An Update on FirstNet*, Testimony by Dr. Damon Darsey, 115th Cong., 1st sess., July 20, 2017, https://www.commerce.senate.gov/public/_cache/files/1dcae52b-4f87-427b-a57a-d6541f680230/62666C224843330B4332AF4E21A45C85.darsey-testimony.pdf.

features such as Proximity Services (ProSe), direct device-to-device communications, and MCPTT standards.[69]

Congress may continue monitoring progress of FirstNet devices to ensure that devices, and features on those devices, are standards-based and interoperable. Further, Congress may consider allowing the flexible use of federal grant funds for multiple technologies (e.g., LMR, LTE, dual-use) to allow interoperability across systems (e.g., FirstNet and LMR systems), and to allow public safety agencies flexibility to transition to FirstNet as they are able.

Cybersecurity

The act requires that the network be safe, secure and reliable, and have protections in place against cyber-attacks. FirstNet/AT&T leveraged industry best practices and federal best practices to ensure a secure network for public safety.[70]

AT&T plans to integrate cybersecurity into devices and applications. All devices are to be screened before users can access the network; AT&T plans to screen devices based on AT&T standards; then FirstNet plans to screen devices, based on the federal standards encompassed in the NIST Cyber Security Framework.[71] FirstNet/AT&T plans to provide a highly secure Identify Management and Federated Credential and Access Management system (ICAM)[72] that enables a secure, single sign-on for FirstNet users. FirstNet is working with the Departments of Homeland Security, Justice, and Commerce and the Office of Management and

---

[69] FirstNet Blog, *FirstNet Tech Team Goes to 2017 Mobile World Congress*, https://www.firstnet.gov/newsroom/blog/firstnet-tech-team-goes-2017-mobile-world-congress.

[70] National Association of State Chief Information Officers (NASCIO), "NASCIO Response to Appendix C-10: NPSBN Cybersecurity," 2015, https://www.nascio.org/Portals/0/Advocacy/2015/SpecialNotice-CyberRESPONSE-%20Final.pdf.

[71] NIST Cybersecurity Framework, https://www.nist.gov/cyberframework.

[72] ICAM webpage on FirstNet.com, https://www.firstnet.com/power-of-firstnet/value-of-firstnet/identity-control.

Budget to leverage cyber best practices and to ensure the network is secure.[73]

FirstNet is to be built with layers of security, designed into RANs, the core network, and service platforms, as well as the devices. Firewalls are to enforce stringent security policies developed in cooperation with DHS and Department of Defense (DoD) to meet NIST requirements. The FirstNet design is to be guided by 3GPP standards for encryption and other standards-based security measures and best practices. Communications is to run on a separate core that is both wireless and encrypted, to ensure public safety communications are secure. Further, AT&T is to provide an operations center to monitor traffic on the FirstNet network and ensure it is secure.

Despite these activities, there are cybersecurity risks. Applications developed for the network by outside developers will need to be screened and secure. A 2017 pilot project conducted by the DHS Science and Technology (S&T) Directorate found that 33 popular apps used by public safety had security and privacy concerns; DHS emphasized the need to screen and test public safety apps for vulnerabilities.[74]

In its July 2017 report, GAO reported that a local government official expressed uncertainty with the security of the FirstNet network, noting that having a large concentration of sensitive public-safety information traveling through one network may be viewed as a target for cyber-attacks.[75] The President's National Security Strategy noted the same risk, and emphasized the need to prioritize U.S. efforts, capabilities, and defenses where cyber-attacks could have catastrophic or cascading

---

[73] Meredith Somers, "FirstNet Teams Triage Cyber Security, Rural Coverage as They Build Out Network," *Federal News Radio*, July 20, 2017, https://federalnewsradio.com/technology-main/2017/07/firstnet-teams-triage-cybersecurity-rural-coverage-as-they-build-out-first-responder-network/.

[74] DHS Science and Technology, "DHS S&T Pilot Project Helps Secure First Responder Apps from Cyber Attacks," press release, December 18, 2017, https://www.dhs.gov/science-and-technology/news/2017/12/18/news-release-st-pilot-project-helps-secure-first-responder.

[75] U.S. Government Accountability Office, Public-Safety Broadband Network: FirstNet Has Made Progress Establishing the Network, but Should Address Stakeholder Concerns and Workforce Planning, GAO-17-569, June 2017, https://www.gao.gov/assets/690/685327.pdf.

consequences, including networks used for national security and public safety.[76]

To ensure the network is secure, as required under the act, Congress may encourage FirstNet to continue to engage with federal partners, and state and local Chief Information Officers (CIO), to ensure cybersecurity policies are in place and enforced. Congress may encourage training on cyber risks for FirstNet users, and information-sharing on cyber incidents.

Applications

Under the act, the FirstNet network was required to evolve as new technologies developed. Bidders to the RFP were required to provide an application ecosystem that supports the nationwide public safety broadband network with services relevant to public safety.[77]

For the past several years, FirstNet has engaged in outreach to public safety to collect input from first responders on useful data applications, and mobile broadband tools and technologies that could save lives. For example, FirstNet allows developers and/or first responders to test apps through the FirstNet lab, and to provide input on new applications.

Both the network and applications on the network need to be interoperable to ensure a consistent user experience, and to ensure features offered through the network do not hinder interoperability. Further, applications developed for FirstNet need to be secure. Applications developed by outside developers could introduce malware, viruses, or other cybersecurity risks to the network. Congress can encourage FirstNet to continue to work with federal agency partners (e.g., DHS, NIST) who bring expertise in cybersecurity, to ensure FirstNet applications are secure, and will not pose a risk to the public safety network or to other federal

---

[76] U.S. President (Trump), "National Security Strategy of the U.S.," https://www.whitehouse.gov/wp-content/uploads/2017/12/NSS-Final-12-18-2017-0905.pdf.

[77] Applications are software programs that run on computers. Each program has a specific "application" for the user, such as fleet management, incident management, management of devices and security settings on devices, cloud storage options. The FirstNet RFP required the development of a broad applications ecosystem for public safety users.

systems. Continual collaboration with federal partners on applications will likely help to ensure that FirstNet applications are secure, interoperable, and will serve future federal users.

Innovation

FirstNet has spurred innovation in public safety communications, providing mobile broadband devices and solutions, innovative applications, and creating opportunities to move public safety toward 5G technologies. Under the Public Safety Innovation Accelerator Program, FirstNet and NIST are conducting public safety broadband research. This program received $300 million under the act to fund research and development in public safety broadband; $38.5 million has been awarded to conduct research on improving resiliency, real-time video analytics, location-based services for public safety, mission critical push-to-talk technologies, and improving indoor communications.[78] Looking ahead, there are three issues of concern:

- It is not clear how the costs of upgrades (to 5G) will be shared between FirstNet/AT&T and state and local public safety agencies. Not all agencies may be able to afford advanced technologies, creating gaps in capabilities across users of the network.
- State and local entities are beginning to address privacy concerns related to video cameras and drones, and jurisdictional concerns related to the siting of small-cell equipment.[79] FirstNet established recommended security and privacy specifications for mobile and Internet of Things (IoT) devices (e.g., traffic cameras, monitoring systems for buildings, sensors which could feed information back

---

[78] NIST, "NIST Awards $38.5 Million to Accelerate Public Safety Communications Technologies," June 13, 2017, https://www.nist.gov/news-events/news/2017/06/nist-awards-385-million-accelerate-public-safety-communications.

[79] CRS Report R42543, *The First Responder Network (FirstNet) and Next-Generation Communications for Public Safety: Issues for Congress*, by Lennard G. Kruger.

to public safety officials through the FirstNet network).[80] While these security and privacy specifications may help to address public concerns about these new technologies, they may not be consistent with state and local laws.

- State and local entities will need to update communication governance bodies, standard operating procedures, training, and exercises to integrate FirstNet technologies into response protocols. Including Chief Information Officers (OCIO) may help with integration of FirstNet technologies and cybersecurity.

Congress may seek to expand eligible activities and allowable costs in federal grants, to enable public safety users, including responders from smaller localities and volunteer responders, to access the advanced technologies offered through FirstNet and AT&T, and to encourage state and local investment in activities that help to integrate FirstNet capabilities into response plans, procedures, training, and exercises.

Affordability and Sustainment

Under the act, the network must be permanently self-funded. The GAO has stated that FirstNet/ AT&T have established a framework to finance the network to meet that requirement.

FirstNet/AT&T can charge users (e.g., public safety users, secondary users) for their use of the FirstNet network. AT&T can also monetize (i.e., earn revenue from) the excess capacity of FirstNet's 20 MHz of broadband spectrum, when it is not in use by public safety.

From this revenue, AT&T must pay (a) FirstNet operating and administrative fees, and has committed (b) $40 billion over the life of the contract to support the build-out, operation, and maintenance of the

[80] Darnell Washington, "First Responders: The Key to Wireless Security?" *Security InfoWatch*, October 16, 2017, http://www.securityinfowatch.com/article/12369845/first-responders-the-key-to-wireless-security.

FirstNet network.[81] Figure 3 provides a graphic of the funding model, as depicted by GAO.

While the act required states to either opt-in or opt-out of deploying the RAN in their state, it did *not* require public safety agencies to subscribe to FirstNet. Even if a state opted into the network (i.e., opted to have FirstNet/AT&T deploy the network in their state), the public safety agencies in the states are not required to subscribe to the network. Agencies can continue to use their existing LMR systems, current commercial cellular/broadband providers, or both.

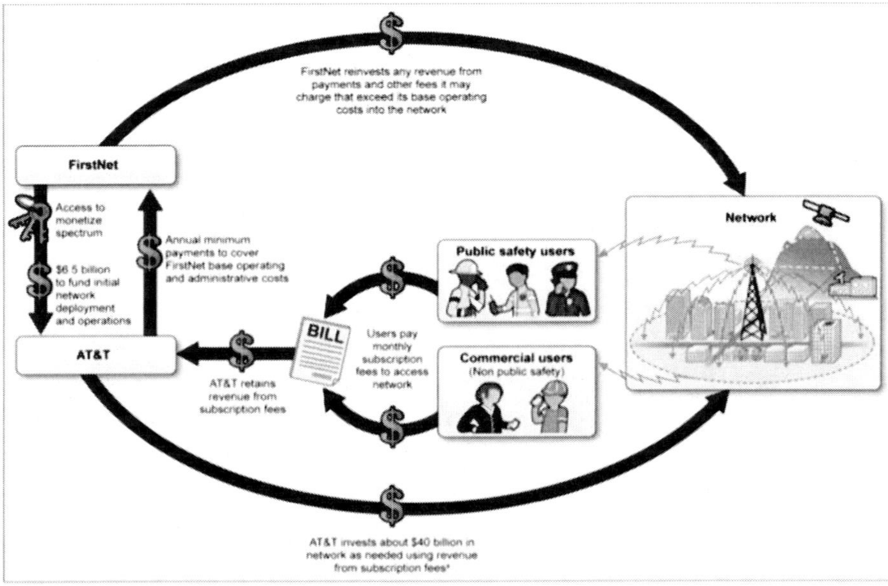

Source: GAO-17-569 U.S. Government Accountability Office, Public-Safety Broadband Network: FirstNet Has Made Progress Establishing the Network, but Should Address Stakeholder Concerns and Workforce Planning, GAO-17-569, June 2017, https://www.gao.gov/assets/690/685327.pdf.

Figure 3. FirstNet Financial Framework.

---

[81] AT&T Corporate webpage on AT&T Public Safety Solutions, https://www.corp.att.com/public-safety/.

A financial challenge for FirstNet/AT&T will be in enrolling subscribers to the system. Likely, FirstNet/AT&T must offer better coverage, services, and features than they currently have, at a price public safety agencies can afford, if they want to attract users to the network. Users will need to weigh the cost of new equipment and new devices, and the cost of the subscription (including any add-on costs for specialized public safety services) in their decision to subscribe.[82]

There are three issues for congressional consideration:

1) FirstNet/AT&T have focused on primary users (e.g., state, local, territory, tribal public safety entities) over the past three years. Congress may ask FirstNet/AT&T about their progress in attracting other users (e.g., federal agencies, 9-1-1 centers, telemedicine providers), to increase subscribers, and to encourage interoperability between agencies, as intended under the law.
2) Many public safety users have stated that they will continue to support LMR (radio) systems for mission critical voice, even if they transition to FirstNet, to ensure first responders always have access to mission critical voice, as primary communications or as back-up communications. User fees for LMR systems have not been able to sufficiently fund state and local LMR systems;[83] adding additional user fees for FirstNet may present financial challenges for state and local agencies.
3) Regional partnerships, added funding streams, and joint purchasing agreements have helped to support LMR systems.[84] Congress may seek to encourage states to develop financial models to support both LMR and FirstNet services. Congress may

---

[82] See California discussion regarding added fees for use of application to access CJIS: https://blog.npstc.org/2018/01/03/californias-firstnet-complaints-highlight-build-out-adoption-challenges/.

[83] State of Minnesota, *Minnesota—Arner and 9-1-1 Funding Study*, February 2014, p. 18, https://dps.mn.gov/divisions/ ecn/programs/armer/Documents/minnesota-armer-and-ng911-funding-study.pdf.

[84] DHS Office of Emergency Communications, *Funding Public Safety Communication Systems*, 2015, https://www.dhs.gov/sites/default/files/publications/Funding%20Mechanisms_Tech Edit_11202015_1.pdf.

consider allowing the flexible use of grant funds to enable state and local agencies to fund LMR systems and FirstNet expenses, including devices and activities that integrate FirstNet capabilities into response, including governance, planning, protocols, training, and exercises. The flexible use of federal grant funds will allow public safety agencies to transition to FirstNet when they are financially and operationally ready.

CONCLUSION

To ensure the $6.5 billion federal funding and 20 MHz of public safety spectrum provided for this project is used to benefit public safety, Congress can monitor the deployment and use of the network over time. Congress may continue oversight of the FirstNet network to better understand the deployment and critical aspects of the network; to ensure that the network is being deployed as intended by Congress; and to ensure that public safety users are subscribing to the network, which is needed to support current operations and future improvements to the network. Congress may include public safety stakeholders in hearings on FirstNet, to ensure that public safety needs (e.g., coverage, resiliency, security) are met and that the network is available, affordable, and sustainable for all users. Congress may inquire about FirstNet's staffing to ensure FirstNet has adequate staff to manage the contract, oversee the deployment, and assist users, as recommended by the GAO.

Lastly, Congress and public safety have long recognized that technology is not the sole driver of an effective response. Congress may seek to encourage state and local agencies to integrate FirstNet capabilities into response plans, protocols, training, and exercises, to ensure that public safety agencies are ready to leverage FirstNet technologies when the next emergency strikes.

APPENDIX A. RELATED ISSUES IN THE ACT

911 Improvements

The act authorized $115 million for a grant program to upgrade public safety answering points (PSAPs) to NG911 capabilities. The NG911 grant program is to provide financial assistance to PSAPs to help fund the establishment of IP-based backbone networks, the application layer software infrastructure needed to interconnect a multitude of response organizations, trainers, call-takers and first responders.

NTIA and NHTSA finalized a management plan to jointly administer the grant and submitted that plan to Congress in January 2017.[85] In September 2017, the NTIA and NHTSA released draft program requirements in the *Federal Register*,[86] seeking comment on the grant. This $115 million grant is to enable some PSAPs to upgrade their networks, but not all PSAPs are to receive funding. Congress can monitor the grant terms and formulas, to ensure that funding is used wisely, and that underserved PSAPs are encouraged to apply.

Congress may seek to ensure that FirstNet and the 911 community engage early, before the deployment of the Band 14 network, to ensure that 911 needs are considered in the earliest stages of network planning and deployment. Congress can encourage FirstNet/AT&T to work in tandem with the 911 community to ensure that upgrades planned for both systems are coordinated, and to ensure that PSAPs can be integrated into the FirstNet network, as encouraged in the act.[87]

---

[85] NHTSA and NTIA, *Management Plan for the Next Generation 9-1-1 Grant Program*, https://www.ntia.doc.gov/files/ntia/publications/nhtsa_ntia_ng911_grant_program_management_plan.pdf.

[86] NHTSA and NTIA, "911 Grant Program," 82 *Federal Register* 44131, September 21, 2017, https://www.federalregister.gov/documents/2017/09/21/2017-19944/911-grant-program.

[87] During 9/11, 9-1-1 call-takers had no information about the impact zone and so were unable to provide informed instructions to callers. The 9/11 Commission, *The 9/11 Commission Report*, July 22, 2004, p. 286, https://www.9- 11commission.gov/report/911Report.pdf.

Spectrum Relocation: T-Band

In the same act that created FirstNet, Congress directed the FCC to reallocate the spectrum in the 470-512 MHz band (also known as the T-Band spectrum) from public safety to commercial use. This spectrum is currently used by 925 public safety licensees in eleven metropolitan areas, for land mobile radio (voice) communications.[88]

The act requires that, within nine years of enactment (i.e., by February 22, 2021), the FCC must reallocate the T-Band public safety spectrum and begin the auction process. Two years after the bidding is complete (~2023), relocation of T-Band public safety entities is required to be completed. NTIA is required to use the proceeds from the bidding process to make grants to cover relocation costs of those public safety entities from the T-Band spectrum.[89]

Public safety advocates believe that Congress saw two options for T-Band operators: relocate those agencies to other frequencies in the area; or, migrate those agencies to the FirstNet network, once it is operational.

Public safety agencies operating on the T-Band argue that there are no frequencies available in the area, which is why they were operating on the T-Band originally. In 2013, NPSTC conducted a study of spectrum and options available to T-Band licensees. The study confirmed probable shortages in several regions, and significant shortages in the five largest urban areas.[90] Further, NPSTC estimated the cost of transition at approximately $5.9 billion, and raised questions as to whether the auction, as it is structured, would be able to attract bidders.[91]

In October 2014, the FCC offered the 700 MHz narrowband reserve channels for general licensing and afforded T-Band public safety licensees

---

[88] Including Boston, MA; Chicago, IL; Dallas/Ft. Worth, TX; Houston, TX; Los Angeles, CA; Miami, FL; New York, NY/Northern NJ; Philadelphia, PA; Pittsburgh, PA; San Francisco/Oakland, CA; Washington, DC/MD/VA.

[89] Section 6103 (b) of P.L. 112-96.

[90] Boston, Chicago, Los Angeles, New York, and Philadelphia.

[91] NPSTC, *T-Band Update Report*, May 31, 2016, p. 31, http://www.npstc.org/download.jsp?tableId=37&column= 217&id=3696&file=T_Band_Update_%20Report_Final.pdf.

priority access to these channels in the T-Band areas.[92] NPSTC issued a *T-Band Update Report* in May 2016, concluding: "Although these 24 additional 700 MHz band channels are certainly beneficial, the number of additional channels pales in comparison to the T-Band channels in use that would need to be relocated to alternative spectrum, especially in the top five T-Band areas."[93]

The second option for T-Band operators suggests agencies could migrate to the FirstNet network. Public safety agencies have argued that this is not a viable option, because FirstNet will not offer mission-critical voice at first launch, which would negatively affect response in those regions.

At a November 1, 2017 hearing of the House Committee on Energy and Commerce, Subcommittee on Communications and Technology, Congress emphasized that it does not want to disrupt critical communications capabilities or lose mission critical voice capabilities during the deployment of the FirstNet network.[94] FirstNet reported that the standards process for MCPTT is proceeding, and it is eager to provide MCPTT to FirstNet primary users. However, the critical factor is time. At its June 2017 Board meeting, FirstNet staff presented board members with a roadmap that includes an update on the standards development process. FirstNet provided a commitment to support MCPTT across the nationwide network by 2019, which brings T-Band agencies closer to the statutory deadline for relocation (~2023) and does not leave a lot of time to transition systems and train people.[95]

---

[92] FirstNet, T-Band Fact Sheet, https://www.firstnet.gov/sites/default/files/T-Band_FactSheet_July2016_0.pdf.

[93] NPSTC, *T-Band Update Report*, May 31, 2016, p. 31, http://www.npstc.org/download.jsp?tableId=37&column=217&id=3696&file=T_Band_Update_%20Report_Final.pdf.

[94] U.S. Congress, Senate Committee on Commerce, Science, and Transportation, Communications, Technology, Innovation, and the Internet, *An Update on FirstNet*, 115th Cong., 1st sess., July 20, 2017; and U.S. Congress, Senate Committee on Commerce, Science, and Transportation, Communications, Technology, Innovation, and the Internet, *Oversight of FirstNet: State Perspectives*, 115th Cong., 1st sess., November 1, 2017. https://energycommerce.house.gov/ hearings/oversight-firstnet-state-perspectives/.

[95] FirstNet Board Minutes, June 28, https://www.firstnet.gov/sites/default/files/Board%20Committee %20Meeting%20June%202017%20.pdf.

The agencies affected by this issue are all located in large population centers. The systems affected by this policy are public safety systems, protecting the life and property in these regions, including the lives of first responders. Congress may consult with affected jurisdictions/agencies to understand the impact of T-Band migration on public safety communications, and to consult with the FCC on relocation options and timelines.

APPENDIX B. RELATED LEGISLATION

Don't Break Up the T-Band Act (H.R. 5085)

Section 6103 of P.L. 112-96 requires public safety agencies to "give back" their T-Band spectrum by 2023. The T-Band (470-512 MHz) is used by public safety agencies in 11 metropolitan regions, for public safety communications. Public safety agencies have argued that there is not sufficient or comparable spectrum for them to migrate, and that the migration would disrupt public safety communications in these populous regions. In February 2018, Representatives Engel, Zeldin, and King of New York introduced the Don't Break Up the T-Band Act of 2018 which calls for the repeal of Section 6103 of P.L. 112-96.

Internet of Things Cybersecurity Improvement Act of 2017 (S. 1691)

In August 2017, Senators Warner, Garner, Wyden, and Daines introduced the Internet of Things (IoT) Act which provides minimum cybersecurity standards for internet-connected devices purchased by federal agencies. This act would direct the Office of Management and Budget (OMB) Director to work with the DoD, General Services Administration (GSA), Department of Commerce, and DHS to issue guidelines for executive agencies to include certain language in contracts for the acquisition of internet-connected devices. This bill would require contractors to certify that hardware, software, or firmware does not have any known vulnerabilities or security defects, to protect federal IT systems. "Under the terms of the bill, vendors who supply the U.S. government with IoT devices would have to ensure that their devices are patchable, do not include hard-coded passwords that can't be changed, and are free of known security vulnerabilities, among other basic requirements."[96] The bill also would require executive agencies to inventory all internet-connected devices in use by the agency, and to create standards for purchases to protect federal systems from malicious attacks. FirstNet and AT&T are already discussing the potential intersection of IoT and public safety (e.g., location data from third-party sources, sensors to monitor infrastructure, video systems to increase situational awareness).

NG911 Act (S. 2061)

In March 2017, Senators Nelson and Klobuchar introduced the NG911 Act which would accelerate the deployment of NG911 services. The NG911 bill would expand a federal grant program to support NG911 improvements. The bill would support standards-based purchases and promote interoperability with FirstNet.

Further, the bill would expand the definition of NG911 to include multiple technologies and multiple means of contacting 9-1-1 (e.g., call, text). This act would also require NIST to prepare a report on NG911 cyber vulnerabilities, the FCC to prepare a report to determine whether agencies are following 9-1-1 best practices, and the GAO to prepare a report on PSAP resiliency and how that can be improved.

---

[96] Senator Mark R. Warner, "Senators Introduce BiPartisan Legislation to Improve Cybersecurity of Internet of Things" (IoT) Devices," press release, August 1, 2017, https://www.warner.senate.gov/public/index.cfm/2017/8/senators-introduce-bipartisan-legislation-to-improve-cybersecurity-of-internet-of-things-iot-devices.

In: Emergency Management
Editor: Stephanie Padbury
ISBN: 978-1-53614-101-6

Chapter 4

EDUCATION-RELATED REGULATORY FLEXIBILITIES, WAIVERS, AND FEDERAL ASSISTANCE IN RESPONSE TO DISASTERS AND NATIONAL EMERGENCIES*

***Boris Granovskiy*[1] *and Alexandra Hegji*[2]**
[1]Education Policy
[2]Social Policy

ABSTRACT

The 21st century has seen the operation of elementary, secondary, and postsecondary educational institutions and the education of the students they enroll disrupted by natural disasters, such as hurricanes and floods, and by national emergencies, such as the terrorist attacks of September 11, 2001. This chapter is intended to inform Congress of existing statutory and regulatory provisions that may aid in responding to future disasters and national emergencies that may affect the provision of

---

* This is an edited, augmented and reformatted version of a Congressional Research Service report, R42881, dated February 28, 2018.

or access to education and highlight the actions of previous Congresses to provide additional recovery assistance. This chapter reflects the most recently available guidance, as of the date of publication, related to education-related disaster response and flexibilities. It will be updated should new or additional statutes be enacted or guidance be issued.

The majority of federal aid for disaster management is made available from the Federal Emergency Management Agency (FEMA) under the authority of the Robert T. Stafford Disaster Relief and Emergency Assistance Act (Stafford Act; P.L. 93-288). Under the Stafford Act, public school districts, charter schools, private nonprofit educational institutions, public institutions of higher education (IHEs), and federally recognized Indian tribal governments are eligible to receive assistance for activities such as debris removal, infrastructure and equipment repair and replacement, hazard mitigation, and temporary facilities.

In addition to the assistance available through the Stafford Act, assistance is available through numerous provisions in education laws. At the elementary and secondary level, there are several existing provisions that may be helpful in providing assistance in response to a disaster. The Elementary and Secondary Education Act (ESEA) grants the Secretary of Education (the Secretary) authority to issue waivers of any statutory or regulatory requirement of the ESEA for a state educational agency (SEA), local educational agency (LEA), Indian tribe, or school that receives funds under an ESEA program and requests a waiver. In response to past disasters, waivers have been granted to address funding flexibility issues and accountability requirements. The Individuals with Disabilities Education Act (IDEA) grants the Secretary authority to waive state maintenance of effort (MOE) requirements and requirements to supplement, not supplant, federal funds under certain circumstances. The Carl D. Perkins Career and Technical Education Act grants the Secretary authority to waive certain accountability metrics and provides some flexibility with regard to MOE requirements. The Secretary is not, however, able to waive all statutory and regulatory requirements with respect to the acts. Under the ESEA, for example, the Secretary may not waive civil rights requirements or prohibitions against the use of funds for religious worship or instruction. Under IDEA, for example, the Secretary may not grant waivers from the right to a free appropriate public education.

At the postsecondary level, various provisions exist to ensure continuity of operations and continuity of federal funding following a disaster. Under the Higher Education Act (HEA), the Secretary of Education has authority to waive several of the requirements for aid recipients, IHEs, and financial institutions when a disaster has been declared. In particular, waivers have been provided from various requirements related to the disbursement, repayment, and administration of federal student aid. Under Title 38 of the U.S. Code, the Department of Veterans Affairs (VA) may extend payment of veteran's educational assistance benefits to cover periods when enrollment is interrupted.

In addition, various provisions exist to provide flexibilities in other Department of Education-administered programs. The Adult Education and Family Literacy Act (AEFLA) and accompanying regulations grant the Secretary the authority to waive state MOE and reporting requirements for AEFLA state grants due to uncontrollable circumstances such as natural disasters. The Rehabilitation Act of 1973, as amended, and accompanying regulations grant the Secretary the authority to waive state MOE and reporting requirements for state vocational rehabilitation grants due to uncontrollable circumstances, such as natural disasters.

In response to the multiple hurricanes and tropical storms in 2017, Congress enacted the Hurricanes Harvey, Irma, and Maria Education Relief Act of 2017, P.L. 115-64, which amended the ESEA to make private schools eligible for funds under the Project School Emergency Response to Violence Program, which provides funds to help schools recover from violent or traumatic events in which the learning environment has been disrupted. The act also provided short-term authority for the Secretary to waive or modify statutory or regulatory requirements to ensure funds were targeted to affected populations and institutions at the elementary, secondary, and postsecondary levels. On February 9, 2018, the Further Additional Supplemental Appropriations for Disaster Relief Act, 2018 (Division B, Subdivision 1 of the Bipartisan Budget Act of 2018, P.L. 115-123) was enacted. It includes FY2018 supplemental appropriations for disaster relief for, among other things, education-related programs and activities for "covered disasters or emergencies," which include the aforementioned hurricanes and tropical storms and also the California wildfires of 2017.

INTRODUCTION

Since 2001, hundreds of thousands of elementary, secondary, and postsecondary education students have been adversely affected by natural disasters, such as hurricanes and floods, and by national emergencies, such as the September 11, 2001, terrorist attacks. To assist Congress in responding to catastrophic events, this chapter provides a general overview of existing statutory and regulatory authorities, flexibilities, and programs that are available to assist students and educational institutions affected by a major disaster or a national emergency. The chapter also reviews several no-longer-authorized temporary provisions that were enacted by Congress to provide additional assistance to support education-related disaster recovery.

Some of the initial education-related problems that may need to be addressed following a disaster relate to utilities outages, displaced students, homeless students, damage to buildings and school property, and school closures. To address such problems, the federal government has responded in various ways to disasters that have affected the provision of education. These responses have included the provision of financial support, waivers of existing statutory and regulatory provisions, and the creation of new programs.

The majority of federal assistance for disaster management is made available from the Federal Emergency Management Agency (FEMA), as authorized under the Robert T. Stafford Disaster Relief and Emergency Assistance Act (Stafford Act: P.L. 93-288).[1] The public assistance available under the Stafford Act largely provides funds for disaster relief activities such as debris removal, emergency protective measures, and the repair, replacement, or restoration of public facilities damaged during the disaster. On several occasions, Congress has also enacted legislation to create temporary targeted assistance programs to support Department of Education (ED) administered programs at the elementary, secondary, and postsecondary education levels. Such assistance was authorized to meet the needs of students, schools, and states during specific disasters. For example, temporary programs have provided funds to restart school operations, reopen schools, and re-enroll students.

In addition to the assistance authorized by the Stafford Act and temporary targeted assistance programs, a wide range of disaster aid is provided by ED under statutory authority that specifically refers to disaster assistance, as well as under general assistance provisions.[2]

---

[1] For more information on FEMA disaster assistance and the Stafford Act, see CRS Report R43784, *FEMA's Disaster Declaration Process: A Primer*, by Bruce R. Lindsay; and CRS Report R43990, *FEMA's Public Assistance Grant Program: Background and Considerations for Congress*, by Jared T. Brown and Daniel J. Richardson.

[2] CRS Report RL31734, *Federal Disaster Assistance Response and Recovery Programs: Brief Summaries*, by Carolyn V. Torsell and Jared C. Nagel.

At the elementary and secondary education level, schools, local educational agencies (LEAs), and states have needed assistance in complying with various educational accountability requirements included in the Elementary and Secondary Education Act (ESEA) and in meeting fiscal requirements. A number of statutory flexibilities can be triggered to provide adjustment to, and in some instances exemptions from, such requirements during periods of disaster.

At the postsecondary education level, institutions of higher education (IHEs) have needed assistance complying with the statutory requirements pertaining to the disbursement and reimbursement of federal student aid (e.g., Pell Grants and Direct Loan program loans), and students have needed assistance in securing, maintaining eligibility for, and repaying federal student aid. Where federal student aid is concerned, relief has been provided related to the interruption or cancelation of periods of study and ensuing challenges associated with reimbursement of funds for the federal government from IHEs and/or students and portability of aid should students need to be relocated.

The first part of this chapter briefly describes recent education-related developments in response to numerous disasters in 2017. It then describes education-related federal assistance provided by FEMA in the aftermath of disasters generally. This is followed by an examination of education-related disaster recovery and issues related to elementary and secondary education, postsecondary education, and various ED-administered programs typically serving adult learners. The chapter also includes an appendix summarizing education-related disaster appropriations administered by ED from between 2005 and 2010.

RECENT DEVELOPMENTS

In August and September of 2017, the United States experienced several major hurricanes and tropical storms in quick succession. In response to Hurricane and Tropical Storm Harvey, President Trump issued a major disaster declaration for Texas[3] and an emergency declaration for Louisiana[4]; in response to Hurricane Irma, President Trump issued emergency declarations for Florida,[5] the U.S. Virgin Islands,[6] and Puerto Rico;[7] and in response to Hurricane Maria, President Trump issued major disaster declarations for Puerto Rico[8] and the U.S. Virgin Islands.[9] In December 2017, California experienced multiple wildfire incidents.[10] The scope of the adverse effects from these natural disasters is extensive. Included among the currently known and expected adverse effects are physical damage to local elementary, secondary, and postsecondary schools, possible school closures, and temporary or permanent displacement of students.[11]

As of the date of this chapter, legislation has been enacted in response to the 2017 hurricanes to provide the Secretary of Education (the Secretary) with waiver authority for certain statutory provisions not otherwise authorized, which relates to the Higher Education Act of 1965 (HEA; P.L. 89-329) Title IV campus-based student aid programs, and to permanently make private schools eligible for funding under the Project

---

[3] Federal Emergency Management Agency, "Texas Hurricane Harvey (DR-4332)," https://www.fema.gov/disaster/ 4332.

[4] Federal Emergency Management Agency, "Louisiana Tropical Storm Harvey (EM-3382)," https://www.fema.gov/ disaster/3382.

[5] Federal Emergency Management Agency, "Florida Hurricane Irma (EM-3385)," https://www.fema.gov/disaster/3385.

[6] Federal Emergency Management Agency, "U.S. Virgin Islands Hurricane Irma (EM-3383), https://www.fema.gov/ disaster/3383.

[7] Federal Emergency Management Agency, "Puerto Rico Hurricane Irma (EM-3384)," https://www.fema.gov/disaster/ 3384.

[8] Federal Emergency Management Agency, "Puerto Rico Hurricane Maria (DR-4339)," https://www.fema.gov/disaster/ 4339.

[9] Federal Emergency Management Agency, "U.S. Virgin Islands Hurricane Maria (DR-4340)," https://www.fema.gov/ disaster/4340.

[10] Federal Emergency Management Agency, "Disasters," https://www.fema.gov/disasters.

[11] See, for example, Lea Wilson and Jake Reiner, "How aftermath of Hurricane Harvey greatly impacts area schools," *KRPC Houston*, September 6, 2017, https://www.click2houston.com/ news/how-aftermath-of-hurricane-harvey-greatly-impacts-area-schools; Ariana Figueroa, "U.S. Schools Brace for an Influx of Students from Puerto Rico," *National Public Radio*, September 27, 2017, http://www.npr.org/sections/ed/2017/09/27/553714097/u-s-schools-brace-for-an-influx-of-students-from-puerto-rico; and David Begnaud, "Many students leaving Puerto Rico as schools struggle without power," *CBS News*, November 19, 2017.

School Emergency Response to Violence Program (Project SERV),[12] as authorized under the ESEA.[13] In addition, ED has released updated guidance for elementary and secondary schools, for IHEs and students related to the federal student aid programs authorized under Title IV of the Higher Education Act, and for other programs administered by ED, including general guidance on grants administration.[14] Finally, ED awarded Project SERV Grants to various state and territory departments of education to assist with hurricane recovery efforts.[15]

On February 9, 2018, the Further Additional Supplemental Appropriations for Disaster Relief Act, 2018 (Division B, Subdivision 1 of the Bipartisan Budget Act of 2018; P.L. 115-123) was enacted. It includes FY2018 supplemental appropriations for disaster relief for "covered disasters or emergencies" (i.e., Hurricanes Harvey, Irma, Maria, or wildfires in 2017 for which a major disaster or emergency has been declared under Sections 401 or 501 of the Robert T. Stafford Disasters Relief and Emergency Assistance Act), including for education-related programs and activities. The act includes the following education-related disaster relief provisions:

- $2.5 billion for immediate aid to restart school operations and for temporary emergency impact aid for displaced K-12 students. These programs are identical to disaster relief programs passed in response to the 2005 hurricane season and authorized under P.L.

---

[12] ESEA §4631. Project SERV funds can be used by LEAs and IHEs to help them and their students recover from violent or traumatic events in which the learning environment has been disrupted.

[13] P.L. 115-56, Hurricanes Harvey, Irma, and Maria Education Relief Act of 2017.

[14] U.S. Department of Education, "Non-Regulatory Guidance on Flexibility and Waivers for Grantees and Program Participants Impacted by Federally Declared Disasters," September 2017, https://www2.ed.gov/policy/gen/guid/ disasters/disaster-guidance.docx; and Kathleen A. Smith, Acting Assistant Secretary for Postsecondary Education, *Guidance for helping Title IV participants affected by a major disaster*, U.S. Department of Education, DCL ID GEN-17-08, Guidance for helping Title IV participants affected by a major disasters, August 29, 2017, https://ifap.ed.gov/dpcletters/GEN1708.html.

[15] U.S. Department of Education, "US Department of Education Awards $8 Million in Project SERV Grants to Hurricane-Impacted States, Territories," press release, November 9, 2017, https://content.govdelivery.com/accounts/ USED/bulletins/1c385ba.

109-148.[16] Funds are targeted to schools that have been damaged by a covered disaster or emergency and to states and LEAs that have enrolled students displaced as a result of a covered disaster or emergency. Funds under the temporary emergency impact aid program are distributed to the states and LEAs based on the numbers of displaced students enrolled. Extra funds are provided to schools for displaced students with disabilities and displaced students who are English Learners.

- Up to $35 million for the Project SERV program, authorized under Section 4631(b) of the ESEA, for education-related services to help students in affected areas recover from the trauma associated with natural disasters.
- $25 million for assistance to LEAs serving homeless children and youths displaced by a covered disaster or emergency, consistent with the Education for Homeless Children and Youth program authorized under Section 723 of the McKinney-Vento Homeless Assistance Act.[17]
- $100 million for the Federal Supplemental Educational Opportunity Grant (FSEOG) program, the Federal Work-Study (FWS) program,[18] and the Fund for the Improvement of Postsecondary Education (FIPSE). Funds shall be for IHEs located in areas affected by a covered disaster or emergency and students enrolled in such institutions. Matching requirements under FSEOG and FWS are waived for IHEs receiving such funds. Funds may be used for a variety of expenses including student financial assistance, staff salaries, student supplies and equipment, or any other purpose authorized under the HEA.

---

[16] See sections titled *Immediate Aid to Restart School Operations* and *Temporary Emergency Impact Aid* in the Appendix of this report for a description of the original programs.

[17] 20 U.S.C. 11431-11435. For more information on the McKinney-Vento Homeless Assistance Act, see CRS Report RL30442, *Homelessness: Targeted Federal Programs*, coordinated by Libby Perl.

[18] For more information on the FWS and FSEOF programs, see CRS Report RL31618, *Campus-Based Student Financial Aid Programs Under the Higher Education Act*, by Joselynn H. Fountain.

- Up to $75 million for IHEs to help defray the unexpected expenses associated with enrolling students displaced from IHEs at which operations have been disrupted by a covered disaster or emergency.
- Authorization for the Secretary to waive, modify, or provide extensions for any statutory or regulatory provisions applicable to the HEA Title IV programs or student eligibility or institutional eligibility provisions in the HEA for affected individuals, students, and IHEs in covered disaster or emergency areas, if the Secretary deems such waiver, modification, or extension necessary in connection with a covered disaster or emergency.[19]
- Authorization for the Secretary to modify required and allowable uses of funds under the various minority-serving grant programs authorized under HEA Title III, Parts A and B, and under the TRIO and GEAR-UP programs.
- Authorization for the Secretary to forgive any outstanding balance owed to ED under the HBCU Hurricane Supplemental Loan program established under P.L. 109-234 and in response to the 2005 hurricane season.[20]
- $4 million to ED's Office of the Inspector General for carrying out administrative and oversight activities.

This chapter reflects the most recently available information on laws and guidance related to education-related disaster response and flexibilities. It will be updated should new or additional statutes be enacted or guidance be issued.

---

[19] Authorized flexibilities mirror those authorized following the 2005 hurricane season under the Higher Education Hurricane Relief Act of 2005 (P.L. 109-148).

[20] Under this provision, four HBCUs—Dillard University, Southern University at New Orleans, Xavier University, and Tougaloo College—are eligible to have their loans forgiven. It is reported that these four schools borrowed a total of $361 million in loans capped at a 1% interest rate to be repaid over 30 years. They have repaid approximately $12.4 million of the loans since borrowing them. Melissa Korn, "Loan Program for Black Colleges Struggles With Oversight, Repayment; Some of the $2 billion in loans ends up more a burden than a help to the schools," *Wall Street Journal*, July 17, 2017.

FEMA AND THE STAFFORD ACT

The principal authority governing federal assistance for emergencies and major disasters in the United States is the Robert T. Stafford Relief and Emergency Assistance Act (Stafford Act; P.L. 93-288, 42 U.S.C. 5121 et seq.). The Federal Emergency Management Agency of the Department of Homeland Security (DHS) has been given the responsibility of administering almost all of the President's Stafford Act authorities through other law, a series of Executive Orders, and a DHS delegation.[21] If a major disaster or emergency has been declared by the President under the authority of the Stafford Act,[22] federal assistance can be made available to the declared tribal governments, states, local governments, owners of certain private nonprofit facilities, and individuals and families.

If located in a declared disaster area, public educational institutions, and certain private nonprofit institutions, may receive disaster assistance grants through FEMA's Public Assistance (PA) Program.[23] PA grants may assist educational institutions with a range of disaster assistance needs, including, but not limited to, the repair and reconstruction of damaged facilities, debris removal, the provision of temporary school facilities, and the replacement of certain necessary equipment and supplies (e.g., books and desks).[24] The specific eligibility of private nonprofit educational facilities is outlined in FEMA policy, but generally includes accredited

---

[21] See 6 U.S.C. §314, Executive Order 12148, 44 FR 43239 (1979), as amended most recently by Executive Order 13286, 68 FR 10619 (2003), and DHS Delegation 9001.1. The President exclusively retains the authority to declare a major disaster or emergency under the Stafford Act §401, 42 U.S.C. §5170, and the Stafford Act §501, 42 U.S.C. §5191, Section 501 of the Stafford Act, respectively.

[22] For information on the Stafford Act disaster declaration process, see CRS Report R43784, *FEMA's Disaster Declaration Process: A Primer*, by Bruce R. Lindsay.

[23] For information on the PA Program, see CRS Report R43990, *FEMA's Public Assistance Grant Program: Background and Considerations for Congress*, by Jared T. Brown and Daniel J. Richardson.

[24] FEMA is prohibited from approving funds for work that is covered by any other source of funding, including funding provided via insurance proceeds and other federal or state agencies. However, FEMA funds may supplement assistance from other sources such that they may be used to cover the remainder of unmet costs for a particular activity. The Stafford Act §312,42 U.S.C. §5155.

primary, secondary, and higher education institutions.[25] Eligible institutions may also apply for funding to reduce risks and damages that might occur in future disasters through hazard mitigation grants.[26]

ELEMENTARY AND SECONDARY EDUCATION

The federal government has provided support for elementary and secondary education for many decades. Federal support for public elementary and secondary education accounts for less than 10% of the nation's overall K-12 education revenues.[27] For FY2017, about $38 billion was appropriated for elementary and secondary education programs administered by ED.[28] The majority of these funds are provided under the Elementary and Secondary Education Act, the Individuals with Disabilities Education Act (IDEA), and the Carl D. Perkins Career and Technical Education Improvement Act (Perkins).

The ESEA represents the major federal commitment to the nation's elementary and secondary schools and was last comprehensively

---

[25] Prior to revisions made to the Stafford Act in the Post-Katrina Emergency Management Reform Act (P.L. 109-295, Title VI), private nonprofit education institutions (but not public institutions) were first required to apply for assistance from the Small Business Administration before applying for and receiving any repair and reconstruction grants under the PA Program. P.L. 109-295 §689h. For additional guidance on the eligibility of private nonprofit educational facilities, see Federal Emergency Management Agency, *Public Assistance Program and Policy Guide*, FP-104-009-2, January 2016, pp. 9-18, at https://www.fema.gov/media-library/assets/documents/111781.

[26] The Stafford Act §404, 42 U.S.C. §5170c. In addition, the Stafford Act allows the heads of federal agencies, such as the Secretary of Education, to modify or waive certain administrative conditions for receiving existing grant funding from their agencies if state or local authorities requesting the waiver are unable to meet those conditions as a result of a major disaster. The Stafford Act §301, 42 U.S.C. §5141.

[27] U.S. Department of Education, National Center for Education Statistics, *Revenues and Expenditures for Public Elementary and Secondary Education: School Year 2012-13 (Fiscal Year 2013)*, January 2016, p. 5, http://nces.ed.gov/pubs2015/2015301.pdf.

[28] This includes funding for career and technical education. The majority of the funds for career and technical education are made available for formula grants to states. States have the discretion to determine how much funding will be directed to elementary and secondary education versus higher education.

reauthorized by the ESSA on December 10, 2015.[29] The Title I-A program is the largest grant program authorized under the ESEA and is funded at $15.5 billion for FY2017. Title I-A of the ESEA authorizes aid to LEAs for the education of disadvantaged children. Title I-A grants provide supplementary educational and related services to low-achieving and other students attending prekindergarten through grade 12 schools with relatively high concentrations of students from low-income families. Title I-A has also become a vehicle to which a number of requirements affecting broad aspects of public K-12 education for all students have been attached as conditions for receiving Title I-A grants. Other ESEA programs include Improving Teacher Quality, 21st Century Community Learning Centers, Rural Education, English Language Acquisition Grants, Charter Schools Grants, and Impact Aid, collectively funded at $8.4 billion in FY2017.

In addition to the ESEA, another important source of federal funding for elementary and secondary education is the IDEA. The act provides federal funding for special education and related services for children with disabilities and requires the provision of a free appropriate public education (FAPE)[30] as a condition for the receipt of such funds. The statute also contains detailed due process provisions to ensure the provision of FAPE and fiscal accountability provisions. Total IDEA funding for FY2017 equaled $13.0 billion.

The Perkins Act is a federal law supporting the development of career and technical education skills among students in secondary and postsecondary education. The bulk of Perkins funds are distributed via a formula to the states, which then distribute funds to local recipients, such as LEAs and community colleges. Perkins was most recently

---

[29] For more information on the ESSA, see CRS Report R44297, *Reauthorization of the Elementary and Secondary Education Act: Highlights of the Every Student Succeeds Act*, by Rebecca R. Skinner and Jeffrey J. Kuenzi. While the ESEA has been comprehensively reauthorized, many of the provisions of the act, including those related to Title I-A accountability, did not take effect immediately. Some provisions, such as new funding formulas for formula grant programs took effect on July 1, 2017, per the Consolidated Appropriations Act of 2016 (P.L. 114-113). Other provisions, such as those related to accountability, are going into effect no earlier than the 2017-2018 school year.

[30] For more information on IDEA and FAPE, see CRS Report R41833, *The Individuals with Disabilities Education Act (IDEA), Part B: Key Statutory and Regulatory Provisions*, by Kyrie E. Dragoo.

comprehensively reauthorized in 2006, and appropriations were authorized through FY2012. The authorization of appropriations was automatically extended for an additional fiscal year through FY2013 under Section 422 of the General Education Provisions Act, and Perkins has continued to receive fairly constant appropriations through FY2016. Total Perkins funding for FY2017 was $1.1 billion.[31]

This section of the chapter provides an overview of existing general statutory and regulatory authorities for elementary and secondary education that enable the Secretary of Education (the Secretary) to waive or modify various education requirements. This section also discusses temporary authorities that have been provided to the Secretary by Congress in response to various disasters. Additionally, it examines temporary elementary and secondary education programs that were created by Congress in response to various disasters and administered by ED. The section concludes with a discussion of other actions taken in response to disasters affecting elementary and secondary education. With respect to elementary and secondary education, the education-related disaster recovery efforts examined in this chapter are primarily related to the 2005 Gulf Coast hurricanes (Katrina and Rita), 2008 Hurricanes Ike and Gustav, and natural disasters that occurred through 2010, including flooding in the Midwest in 2008. ED has not issued waivers or new funding related to disasters that occurred subsequent to this time period.

While the ESEA was recently reauthorized by the ESSA, this chapter discusses relevant ESEA provisions both prior to and following the enactment of the ESSA.

Current Waiver Authority and Flexibility

Existing statutory and regulatory requirements provide the Secretary with some authority and flexibility to respond to a disaster without the

---

[31] For more information about Career and Technical Education and the Perkins Act, see CRS Report R44542, *Carl D. Perkins Career and Technical Education Act of 2006: An Overview*, by Boris Granovskiy.

need for additional legislation. This section begins with an examination of waiver authority available to the Secretary under the ESEA. The subsequent text illustrates how waivers were used in response to recent disasters to address funding flexibility issues and accountability requirements under the ESEA. Other forms of flexibility that are available to the Secretary are also discussed. The extent to which the Secretary has flexibility under IDEA and Perkins is also described.

Waiver Authority under the ESEA as Amended by the ESSA

Existing statutory and regulatory waiver authorities under ESEA are intended to support state and local educational agencies and schools that receive ESEA funds and are located in areas affected by disasters. Typical waiver requests that have been granted by ED in the past have focused on relief from accountability requirements, such as counting displaced students as a separate subgroup when making annual yearly progress determinations, and on relief from certain fiscal requirements, such as extending the period of availability of various ESEA funds that would have expired by a certain date.

Under the ESEA, as amended by the ESSA, secretarial waiver provisions are included in Section 8401.[32] Under Section 8401, the Secretary has broad authority to issue waivers of a number of statutory or regulatory requirements of the ESEA for a state education agency (SEA); local education agency, through an SEA; an Indian tribe; or a school, through an LEA that receives funds under an ESEA program and requests a waiver.

The Secretary is prohibited from waiving any statutory or regulatory requirement related to the following requirements:

- allocation or distribution of funds to states or LEAs (or other grant recipients);

---

[32] Prior to the enactment of the ESSA, these provisions were included in Section 9401 of the ESEA.

- maintenance of effort (MOE) requirements for LEAs or SEAs to maintain their level of spending for specified educational services;[33]
- comparability of services;[34]
- the use of federal aid only to supplement, and not supplant, state and local funds for specified purposes;[35]
- equitable participation of private school students and teachers;
- parental participation and involvement;
- applicable civil rights requirements;
- the requirement for a charter school under the Public Charter Schools program (Title IV-C);
- prohibitions against consideration of ESEA funds in state aid to LEAs;[36]
- prohibitions against use of funds for religious worship or instruction;[37]
- certain prohibitions against use of funds for sex education;[38] and
- certain ESEA Title I-A school selection requirements.

The Secretary is prohibited from disapproving a waiver request based on conditions outside the scope of such request. Waivers may be approved by the Secretary for a period not to exceed four years, but may be extended

---

[33] To comply with MOE requirements, LEAs must provide, from state and local sources, a level of funding (either aggregate or per pupil) in the preceding year that is at least a specified 90% of the amount provided in the second preceding year. For more information about fiscal accountability requirements, see CRS In Focus IF10405, *Fiscal Accountability Requirements That Apply to Title I-A of the Elementary and Secondary Education Act (ESEA)*, by Rebecca R. Skinner.

[34] For more information about fiscal accountability requirements, see CRS In Focus IF10405, *Fiscal Accountability Requirements That Apply to Title I-A of the Elementary and Secondary Education Act (ESEA)*, by Rebecca R. Skinner.

[35] Federal program funds must be used to supplement, and not supplant, state and local funds that would otherwise be available for the specific services or activities for which federal funds may be used under the program in question. For more information about fiscal accountability requirements, see CRS In Focus IF10405, *Fiscal Accountability Requirements That Apply to Title I-A of the Elementary and Secondary Education Act (ESEA)*, by Rebecca R. Skinner.

[36] ESEA §8522.

[37] ESEA §8505.

[38] ESEA §8526.

if the state demonstrates that the waiver was effective in enabling the state to carry out the activities for which the waiver was requested, the waiver contributed to improved student academic achievement, and extension of the waiver is in the public interest.

Use of Prior ESEA Section 9401 Waiver Authority in Response to Disasters

This portion of the report discusses the use of the ESEA Section 9401 waiver authority, which was the waiver authority in place prior to the enactment of the ESSA. Although no waivers have yet been granted since the enactment of the ESSA, information on waivers granted under the previous Section 9401 may be useful in determining the potential availability of waivers, as the previous Section 9401 waivers were maintained in current Section 8401.

The ESSA maintained the waiver authority previously provided under Section 9401 of the ESEA to the Secretary to approve or deny waivers requested by SEAs, LEAs, Indian tribes, or schools (through an LEA). These statutory requirements did not specify that these requests be made in response to a natural or man-made disaster, financial issues, or other circumstances. While many of the waiver requests that have been approved by the Secretary since enactment of the No Child Left Behind Act of 2001[39] (the comprehensive reauthorization of the ESEA immediately prior to the ESSA) have focused on ESEA Title I-A educational accountability requirements, funding issues, general administrative requirements, and issues related to the American Recovery and Reinvestment Act (ARRA; P.L. 111-5), waivers have also been granted in response to natural disasters. As this waiver authority was maintained under the ESEA as amended by the ESSA, SEAs, LEAs, Indian tribes, or schools interested in obtaining a waiver in response to a natural disaster or other circumstances, such as an incident of school violence, could also submit a waiver request for consideration by the Secretary.

---

[39] P.L. 107-110.

Table 1. Summary of Waivers Granted in Response to Disasters and Other Relevant Issues Pursuant to Section 9401 of the Elementary and Secondary Education Act (ESEA): 2005-2010

State	Number of Waivers Granted by Year	Purpose of Waivers
Alabama	1—2006	Waiver was related to accountability requirements, specifically creating a separate subgroup for displaced students for making adequate yearly progress (AYP) determinations.[a]
Arkansas	1—2006	Waiver was related to accountability requirements, specifically creating a separate subgroup for displaced students for making AYP determinations.[a]
California	1—2010	Waiver permitted Calexico Unified School District to exclude assessments administered during the 2009-2010 school year when making AYP determinations and to assign the same AYP status to schools as it assigned based on 2008-2009 school year assessments as an earthquake and aftershocks resulted in schools closing for a period of six weeks that coincided with the 2009-2010 assessment administration period.
Florida	1—2010	Waiver allowed Florida to exclude from assessment participation rate calculations used to determine AYP for the 2009-2010 school year, students who enrolled in Florida schools after being displaced as a result of the January 12, 2010, earthquake in Haiti.
Georgia	2—2006	Both waivers were related to accountability requirements. The first waiver permitted the state to create a separate subgroup for displaced students for making AYP determinations.[a] The second waiver permitted a local educational agency (LEA) to allow schools to offer supplemental educational services (SES) after failing to make AYP for two consecutive years and offer school choice after failing to make AYP for a third year.[b]
Louisiana	3—2005 5—2006 2—2007 1—2008 2—2009	Waivers were related to both accountability requirements and funding issues: permitting the state to create a separate subgroup for displaced students for making AYP determinations;[a] providing LEAs and schools that were closed for 18 or more days and located in a declared disaster area with greater flexibility in making AYP determinations; permitting LEAs to allow schools to offer SES after failing to make AYP for two consecutive years and offer school choice after failing to make AYP for a third year;b extending the period of availability of various funds that would have expired by a certain date; enabling the state to waive Title I-A fund carryover requirements for its LEAs;c and waiving requirements with respect to the use of funds under the Educational Technology program (Ed Tech; ESEA Title II-D).d

Table 1. (Continued)

State	Number of Waivers Granted by Year	Purpose of Waivers
Maryland	1—2010	Waiver allowed Maryland to exclude schools from assessment participation rate calculations for AYP determinations for the 2009-2010 school year due to student absences during the testing period due to the H1N1 flu virus or other flu-like illnesses.
Mississippi	2—2005 1—2006 2—2007 2—2008	Waivers were related to both accountability requirements and funding issues. With respect to accountability, the first waiver permitted the state to create a separate subgroup for displaced students for making AYP determinations.[a] With respect to funding flexibility, four waivers extended the period of availability of various ESEA funds that would have expired by a certain date. Another waived requirements with respect to the use of funds under Ed Tech.[d] A sixth waiver enabled the state to waive Title I-A fund carryover requirements for its LEAs.[c]
New Mexico	1 - 2010	Waiver allowed New Mexico to exclude schools or LEAs from attendance rate as an "other academic indicator" when making AYP determinations for the 2009-2010 school year for a school or LEA that failed to make AYP for the attendance rate due to the H1N1 flu virus or other flu-like illnesses and use data from the prior school year instead.
South Carolina	1—2010	Waiver allowed South Carolina to exclude schools or LEAs from attendance rate as an "other academic indicator" when making AYP determinations for the 2009-2010 school year for a school or LEA that failed to make AYP for the attendance rate due to the H1N1 flu virus or other flu-like illnesses and use data from the prior school year instead.
Tennessee	1—2006	Waiver was related to accountability requirements, specifically creating a separate subgroup for displaced students for making AYP determinations.a
Texas	2—2006 1—2008	Two of the three waivers were related to accountability requirements. The first waiver permitted the state to create a separate subgroup for displaced students for making AYP determinations.a The second waiver provided seven LEAs that were closed for seven or more days and located in a declared disaster area with greater flexibility in making AYP determinations. The third waiver extended the period of availability of ESEA funds provided under the Reading First program (ESEA Title I-B-1).e

Sources: Table prepared by CRS, August 26, 2016, based on CRS examination of U.S. Department of Education, "Notice of Waivers Granted Under Section 9401 of the Elementary and Secondary Education Act, as Amended," 72 *Federal Register* 10990-10996, March 12, 2007; U.S. Department of Education, "Notice of Waivers Granted Under Section 9401 of the Elementary and Secondary Education Act, as Amended," 73 *Federal Register* 19056- 19058, April 8, 2008; U.S.

Department of Education, "Notice of Waivers Granted Under Section 9401 of the Elementary and Secondary Education Act, as Amended," 74 *Federal Register* 22909-22913, May 15, 2009; U.S. Department of Education, "Notice of Waivers Granted Under Section 9401 of the Elementary and Secondary Education Act, as Amended," 75 *Federal Register* 56833-56856, September 16, 2010; and U.S. Department of Education, "Notice of Waivers Granted the Elementary and Secondary Education Act of 1965, as Amended," 78 *Federal Register* 71410-71429, November 27, 2013; as well as reports submitted to Congress on the waivers granted under Section 9401 that were submitted in April 2007, April 2008, June 2009, and December 2010; available at http://www2.ed.gov/nclb/freedom/local/flexibility/waiverletters/index.html.

Notes: For detailed information about ESEA accountability requirements prior to the enactment of the ESSA, see CRS Report R41533, *Accountability Issues and Reauthorization of the Elementary and Secondary Education Act*, by Rebecca R. Skinner (available upon request to CRS).

[a] Under ESEA prior to the enactment of ESSA, data had to be disaggregated for a number of subgroups to determine whether a school or LEA met AYP. These subgroups included economically disadvantaged students, limited English proficient students, students with disabilities, and students in major racial and ethnic groups as determined by the state.

[b] In the absence of a waiver, the ESEA, prior to the enactment of ESSA, required states to identify LEAs and schools that failed to meet AYP standards for two consecutive years for improvement. Students attending these schools had to be provided with options to attend other public schools that made AYP. If a Title I-A school failed to meet AYP standards for a third year, students from low-income families had to be offered the opportunity to receive instruction from an SES provider of their choice. SES was often provided in the form of afterschool tutoring.

[c] Under ESEA, prior to the enactment of ESSA, LEAs were able to only carry over up to 15% of their Title I-A funds from one fiscal year to another. An SEA was only permitted to waive this requirement once every three years for a given LEA. This provision remained unchanged under ESSA.

[d] Under ESEA, prior to the enactment of ESSA, the Ed Tech program authorized grants to SEAs and LEAs to increase access to educational technology, support the integration of technology into instruction, enhance technological literacy, and support technology-related professional development of teachers. Ed Tech is not authorized under ESSA.

[e] Under ESEA, prior to the enactment of ESSA, the Reading First program was intended to improve reading programs for students in grades K-3 in schools that either had percentages of students from low-income families that were among the highest in the LEA or had been identified for improvement under Title I-A. Reading First is not authorized under ESSA.

As detailed in Table 1, from 2005 through 2010,[40] the Secretary used the authority available under previous ESEA Section 9401 to grant 33 waivers in response to natural disasters.[41] Of these waivers, 18 were

---

[40] Data on waivers provided beyond the 2010 calendar year have not yet been published in the *Federal Register*.

[41] The Secretary used the authority available under ESEA §9401 to grant other types of waivers, such as waivers related to ESEA accountability requirements. For more information about these waivers, see CRS Report R42328, *Educational Accountability and Secretarial Waiver Authority Under Section 9401 of the Elementary and Secondary Education Act*, by Rebecca R. Skinner and Jody Feder (available upon request to CRS).

granted in either 2005 or 2006, 4 were granted in 2007, 4 were granted in 2008, 2 were granted in 2009, and 5 were granted in 2010. Of the waivers granted, just over half of the waivers (17) addressed funding flexibility issues, while the others (16) addressed educational accountability requirements under the ESEA. About 40% of the waivers (13 waivers) were granted to Louisiana, which was disproportionately affected by disasters from 2005 to 2010 compared to other states.

Other Flexibilities Related to Elementary and Secondary Education

Beyond the authority currently available to the Secretary under Section 8401 of the ESEA, other flexibilities that may be exercised by the Secretary, SEAs, or LEAs also exist. Some of these flexibilities are contained in the ESEA, while other flexibility authority is provided through other statutes. None of the flexibility provisions discussed below are specific only to disasters. Rather, these flexibilities may be provided in response to a disaster or for other reasons, such as a "precipitous decline in the financial resources" of an LEA.

Other ESEA Flexibilities

There are several additional provisions included in the ESEA that may be helpful in providing assistance in response to a disaster, including flexibility related to MOE requirements, charter schools, and funding flexibility.

Under the ESEA, the Secretary is specifically prohibited from waiving MOE requirements using authority available under Section 8401. However, Section 8521(c) of the ESEA permits the Secretary to waive MOE requirements for LEAs in the case of "exceptional or uncontrollable circumstances, such as a natural disaster" or due to a "precipitous decline in the financial resources" of an LEA.[42] CRS has been unable to identify examples where this flexibility was used by LEAs.

Under the ESEA, the Charter Schools Program (CSP; ESEA Title IV-C) provides grants to SEAs or, if a state's SEA chooses not to participate,

---

[42] The Secretary may also waive MOE due to a change in the organizational structure of an LEA.

charter school developers, to support the development and initial implementation of public charter schools. CSP also includes general provisions for allocations to charter schools under ESEA and other federal formula grant programs. Under Section 4303(d)(5) of the ESEA, the Secretary has the authority to waive statutory and regulatory requirements if the Secretary determines that granting such a waiver will promote the purpose of the program.[43] Prior to the enactment of the ESSA, the Secretary used this authority to provide funding specifically to support existing charter schools and to create new charter schools in areas affected by disasters.[44]

Additional flexibility under the ESEA allows funds received by LEAs under either the Small, Rural School Achievement Program[45] or the Rural and Low-Income School Program[46] to also be used for activities authorized under several other ESEA programs. Finally, Title V-A provides authority that allows most LEAs and SEAs to transfer up to 100% of their formula grants among various ESEA programs.[47]

Flexibilities Available under ED-Flex

The Education Flexibility Partnership Act of 1999 (ED-Flex; P.L. 106-25) transfers from ED to SEAs the authority to grant waivers similar to those authorized under ESEA Section 8401. The SEAs that receive this authority are similarly restricted in terms of waivers they are not permitted

[43] The Secretary is prohibited from waiving the elements of a charter school included in the definition of a charter school in Section 4310(2).

[44] Department of Education, "Key Policy Letters Signed by the Education Secretary or Deputy Secretary," September 12, 2005, available online at http://www.ed.gov/policy/elsec/guid/secletter/050914.html.

[45] ESEA Title V-B-1.

[46] ESEA Title V-B-2.

[47] The transferability authority permits states to transfer up to 100% of the nonadministrative funds allotted to the state under Title II-A, Title IV-A (block grant program), or 21st Century Community Learning Centers among these programs or to Title I-A, Migrant Education, Neglected and Delinquent, English Language Acquisition, and one other program that is cited in statutory language only as "Part B." Similarly, LEAs can transfer up to 100% of funds under Title II-A or Title IV-A (block grant program) between these programs or to Title I-A, Migrant Education, Neglected and Delinquent, English Language Acquisition, and one other one other program that is cited in statutory language only as "Part B." States and LEAs are prohibited from transferring funds from Title I-A, Migrant Education, Neglected or Delinquent, English Language Acquisition, and one other program that is cited in statutory language only as "Part B" to any other program.

to authorize (e.g., waiving maintenance of effort requirements).[48] Thus, an SEA could grant permitted waivers in response to a disaster or for other reasons without having to seek additional authority from the Secretary. Prior to the enactment of the ESSA, 10 states were authorized to participate in ED-Flex.[49] Under the provisions of the ESSA,[50] these 10 states retain their authority to participate in ED-Flex for up to five years pending approval by the Secretary. In addition, other states may apply to receive ED-Flex authority. Under the ESSA, states authorized to participate in ED-Flex may waive statutory or regulatory provisions under the following ESEA programs:

- Title I-A (except provisions in Section 1111),
- Migrant Education (Title I-C),
- Prevention and Intervention Programs for Children and Youth Who Are Neglected, Delinquent, or At-Risk (Title I-D),
- Preparing, Training, and Recruiting High-Quality Teachers, Principals, and Other School Leaders (Title II-A), and
- Student Support and Academic Enrichment Grants (block grants, Title IV-A). In addition, SEAs may waive statutory or regulatory provisions under the Perkins Act.

Flexibilities Available under IDEA

Under IDEA, the Secretary does not have authority comparable to what is available under ESEA Section 8401. However, the Secretary does have authority to waive MOE requirements for states (but not LEAs) under certain circumstances and, unlike provisions under the ESEA, IDEA provisions allow the Secretary to waive supplement, not supplant requirements for states (but not LEAs) under certain circumstances.

---

[48] In addition, the SEA must demonstrate that the underlying purposes of the statutory requirements of the program for which a waiver is granted will continue to be met "to the satisfaction of the Secretary."

[49] The 10 states are Colorado, Delaware, Kansas, Massachusetts, Maryland, North Carolina, Oregon, Pennsylvania, Texas, and Vermont.

[50] The Education Flexibility Partnership Act of 1999 Reauthorization was included in ESSA Section 9207. It is not considered part of the ESEA.

In general, a state may not reduce the amount of its financial support for special education and related services for children with disabilities below the amount of that support for the preceding fiscal year. In any fiscal year in which a state does not meet this MOE requirement, the Secretary is required to reduce the state's subsequent year grant by the same amount by which the state failed to meet the requirement. The Secretary may grant a waiver for one fiscal year at a time in the case of "exceptional or uncontrollable circumstances" such as a natural disaster or a "precipitous and unforeseen decline in the financial resources of the state." For FY2006, the Secretary used this authority under IDEA Section 612(a)(18)(c) to waive Louisiana's state level MOE requirements. In addition, waivers can be granted if a state can provide "clear and convincing evidence" that FAPE is available for all children with disabilities. If a state does not meet its MOE requirement for any year, including any year for which the state was granted a waiver, the state financial support required in future years is not reduced. That is, the state must provide the amount that would have been required in the absence of failing to meet MOE in the previous year.

Both states and LEAs must use IDEA funds to supplement state, local, and other federal funds and not to supplant them. As with the state MOE requirement, the Secretary has authority to grant a waiver of the state-level supplement, not supplant requirement if the state provides "clear and convincing evidence" that all children with disabilities in the state have FAPE available.

Flexibilities Available under Perkins

Under the Perkins accountability system, a state is required to design and implement a program improvement plan if it does not reach certain targets on its core indicators of performance. If the state fails to attain those improvement plan goals, the Secretary may then withhold a portion of the state's Perkins allotment. The Secretary may waive these sanctions due to "exceptional or uncontrollable circumstances, such as a natural disaster."[51]

---

[51] Perkins §123(a)(3)(B).

Similarly, at the local level, a state can waive sanctions for local recipients of Perkins funds due to exceptional or uncontrollable circumstances.[52]

In addition, the Secretary has some flexibility with respect to the MOE provision in Perkins. The Secretary may waive up to 5% of the MOE requirement for states for one year due to exceptional or uncontrollable circumstances. The level of funding allowed under such a waiver may not be used as a basis for calculating MOE requirements for subsequent years. That is, the state must provide the amount that would have been required in the absence of failing to meet MOE in the previous year.[53]

Temporary Programs Created in Response to Disasters[54]

In response to various natural disasters, Congress has enacted legislation to create temporary programs and give the Secretary temporary authority to issue various waivers and flexibilities at the elementary and secondary education levels to meet the needs of students, schools, LEAs, and states. A brief summary of various programs created in response to disasters is included in the Appendix. For example, temporary programs have provided funds to restart school operations, reopen schools, and re-enroll students. They have also made funds available to recruit, retain, and compensate school staff who have committed to work for a certain number of years in areas affected by disasters. Funds have also been provided to replace instructional materials and equipment; support charter school expansion; modernize, renovate, or repair school buildings; and support extended learning time activities. In response to large numbers of students being displaced by a disaster, a temporary program was created to provide funding to assist both public and nonpublic schools in enrolling displaced students. The decision about whether a student would attend a public or nonpublic school was made by the displaced student's parents.

---

[52] Perkins §123(a)(4)(B).

[53] Perkins §312(b)(2).

[54] The statutory language authorizing these programs is included in P.L. 109-148, P.L. 109-234, P.L. 110-28, and P.L. 111-117.

Most of these temporary programs were authorized in response to Gulf Coast hurricanes between 2005 and 2010. The authorizations of all of these programs have expired. More recently, Congress passed legislation authorizing temporary education-related programs in response to the 2017 natural disasters, including Hurricanes Harvey, Irma, and Maria, and the California wildfires of December 2017.

The Further Additional Supplemental Appropriations for Disaster Relief Act, 2018

In the 115th Congress, the Further Additional Supplemental Appropriations for Disaster Relief Act, 2018 (Division B, Subdivision 1 of the Bipartisan Budget Act of 2018, P.L. 115-123) was enacted. As it relates to elementary and secondary education, the act includes FY2018 supplemental appropriations for disaster relief for "covered disasters or emergencies" (i.e., Hurricanes Harvey, Irma, Maria, or California wildfires in 2017 for which a major disaster or emergency has been declared under Sections 401 or 501 of the Robert T. Stafford Disasters Relief and Emergency Assistance Act). Specifically, the supplemental appropriations for elementary and secondary education include the following:

- $2.5 billion for immediate aid to restart school operations and for temporary emergency impact aid for displaced K-12 students. These programs are identical to disaster relief programs passed in response to the 2005 hurricane season and authorized under P.L. 109-148.[55] Funds are targeted to schools that have been damaged by a covered disaster or emergency and to states and LEAs that have enrolled students displaced as a result of a covered disaster or emergency. Funds under the temporary emergency impact aid program are distributed to the states and LEAs based on the numbers of displaced students enrolled. Extra funds are provided to schools for displaced students with disabilities and displaced students who are English Learners.

---

[55] See sections titled *Immediate Aid to Restart School Operations* and *Temporary Emergency Impact Aid* in the Appendix of this report for a description of the original programs.

- Up to $35 million for the Project SERV Program, authorized under Section 4631(b) of the ESEA, for education-related services to help students in affected areas recover from the trauma associated with natural disasters.
- $25 million for assistance to LEAs serving homeless children and youths displaced by a covered disaster or emergency, consistent with the Education for Homeless Children and Youth program authorized under Section 723 of the McKinney-Vento Homeless Assistance Act.[56]

Postsecondary Education

The federal government supports postsecondary education, primarily through programs authorized by the Higher Education Act of 1965, as amended (HEA; P.L. 89-329). The HEA authorizes a broad array of federal student aid programs that assist students and their families with paying for or financing the costs of obtaining a postsecondary education. In FY2016, nearly $136 billion in grant, loan, and work-study aid was made available to 13 million students enrolled in approximately 6,000 institutions of higher education.[57] The HEA also authorizes a series of programs that provide support to IHEs and additional federal student aid. About $948 million was appropriated for these postsecondary programs in FY2016.[58]

The Department of Veterans Affairs (VA) administers entitlement programs providing vocational and educational assistance. The veterans educational assistance programs (GI Bills®)[59] provide eligible service-members and veterans and their dependents with financial assistance while

---

[56] 20 U.S.C. 11431-11435; for more information on the McKinney-Vento Homeless Assistance Act, see CRS Report RL30442, *Homelessness: Targeted Federal Programs*, coordinated by Libby Perl.

[57] U.S. Department of Education, Office of Federal Student Aid, *Annual Report 2016*, November 14, 2016.

[58] This includes funding for developing institutions, teacher quality partnership programs, campus-based childcare programs for students, and international and foreign language programs.

[59] GI Bill® is a registered trademark of the U.S. Department of Veterans Affairs.

enrolled in approved educational and training programs.[60] The Vocational Rehabilitation and Employment (VR&E) program provides education, job training, and related services to veterans with service-connected disabilities.[61] The VA estimates that it will obligate approximately $15 billion in FY2017 for veterans educational assistance and VR&E.[62]

This section provides an overview of the existing statutory and regulatory authorities that have been and may be useful in supporting education-disaster recovery. The first section examines the existing general statutory and regulatory authorities that enable the Secretary of Education and IHEs to waive or modify various HEA requirements. This is followed by the description of an HEA program that authorizes education disaster and emergency relief loans. The next part of this section examines temporary higher education provisions and programs that were enacted by Congress in response to various disasters and administered by ED. Finally, a discussion of existing general statutory and regulatory authorities that reduce the negative financial impact of disasters on GI Bill participants follows.

Current HEA Waiver Authority and Flexibility

Existing statutory and regulatory provisions grant the Secretary of Education and IHEs a degree of authority and flexibility to respond to a disaster without the need for additional legislation. This section examines waiver authority and other flexibilities available to the Secretary and IHEs under the HEA.

---

[60] For more information on veterans educational assistance programs, see CRS Report R42755, *The Post-9/11 Veterans' Educational Assistance Act of 2008 (Post-9/11 GI Bill): A Primer*, by Cassandria Dortch; and CRS Report R42785, *GI Bills Enacted Prior to 2008 and Related Veterans' Educational Assistance Programs: A Primer*, by Cassandria Dortch.

[61] For more information on the VR&E program, see CRS Report RL34627, *Veterans' Benefits: The Vocational Rehabilitation and Employment Program*, by Benjamin Collins.

[62] U.S. Department of Veterans' Affairs, FY2017 Congressional Budget Justifications.

Regulatory Flexibility and Waivers under the HEA Federal Student Aid Programs

Programs authorized under Title IV of the HEA are the primary source of federal student aid to support postsecondary education. The largest Title IV student aid programs are the Pell Grant program (Title IV-A)[63] and the Direct Loan (DL) program (Title IV-D).[64] Existing statutory and regulatory authorities are intended to support Title IV aid recipients and their families who at the time of the disaster were residing in, employed in, or attending an IHE located in an area designated as a federally declared disaster area as defined in the Stafford Act.[65] Flexibility is also available to IHEs, lenders, servicers, and guaranty agencies that are located in these areas. In addition, ED has indicated that it will provide assistance, on a case-by-case basis, in addressing concerns related to specific regulatory requirements.

Provisions for IHEs

Types of support that may be available to help IHEs in the aftermath of a disaster include the following:

- IHEs may temporarily close and reopen without automatically losing eligibility to participate in the Title IV student aid programs.
- On a case-by-case basis, IHEs may be permitted to shorten the length of an academic year without losing eligibility to participate in the Title IV student aid programs.

---

[63] For a detailed description of the Federal Pell Grant program, see CRS Report R42446, *Federal Pell Grant Program of the Higher Education Act: How the Program Works and Recent Legislative Changes*, by Cassandria Dortch.

[64] For a detailed description of the DL program, see CRS Report R40122, *Federal Student Loans Made under the Federal Family Education Loan Program and the William D. Ford Federal Direct Loan Program: Terms and Conditions for Borrowers*, by David P. Smole.

[65] For a detailed description of the authority, see Kathleen A. Smith, Acting Assistant Secretary for Postsecondary Education, *Guidance for helping Title IV participants affected by a major disaster*, U.S. Department of Education, DCL ID GEN-17-08, Guidance for helping Title IV participants affected by a major disasters, August 29, 2017, https://ifap.ed.gov/dpcletters/GEN1708.html. See also U.S. Department of Education, "Non-Regulatory Guidance on Flexibility and Waivers for Grantees and Program Participants Impacted by Federally Declared Disasters," September 2017, https://www2.ed.gov/policy/gen/guid/disasters/disaster-guidance.docx.

- IHEs may be granted flexibility in providing the required data to ED for its Annual Campus Security Report or its Equity in Athletics Disclosure Report by the established deadlines.
- On a case-by-case basis, IHEs may be provided flexibility in complying with some Title IV institutional financial management requirements, such as cash management.
- On a case-by-case basis, IHEs may be provided flexibility in meeting the filing deadline for the Fiscal Operations Report and Application to Participate (FISAP) for the campus-based programs.
- IHEs may be excepted from requirements related to the records retention in instances where records are lost or destroyed and cannot be reconstructed.
- IHEs may be excepted from requirements for verification of federal student aid (FSA) applicants whose records were lost or destroyed because of a disaster, provided that the IHE has attempted to preserve and reconstruct any applicable records and documents that verification could not be completed due to a disaster.
- IHEs may be granted a waiver of the requirement to obtain signed documentation from the parents of dependent undergraduate students of the number of family household members and the number of family household members enrolled in postsecondary institutions in instances where neither of the parents can provide signed documentation due to the disaster.
- IHEs that are unable to meet deadlines for reporting federal student aid disbursement records or student enrollment status must promptly contact and inform their School Participation Team or the National Student Loan Data System Customer Service Center, respectively, about the situation and in certain circumstances may adjust or modify their reporting schedule.
- Upon request, an IHE may receive an extension of the published deadline for reporting final Federal Pell Grant payments.

- IHEs may be granted a waiver from having their campus-based programs allocations reduced for the second succeeding year due to the under-utilization of funds in the most recent award year, if the under-utilization is solely due to a disaster.[66]
- IHEs may be granted a waiver of either or both the Federal Work-Study program (FWS) requirements that IHEs use at least 7% of their FWS allocation to compensate students employed in community service and that they must have at least one tutoring or family literacy project.[67]

Provisions for Student Aid Recipients

Types of support that may be available to help Title IV student aid recipients in the aftermath of a disaster include the following:

- Students may be permitted to continue their program of education at another IHE while receiving Title IV assistance in accordance with agreements between institutions to permit the continuation of study.
- Any federal or state disaster aid received by a student or the student's family will be excluded from the student's Expected Family Contribution (EFC) and from estimated financial assistance from other sources for purposes of packaging federal student aid.[68]
- An IHE's financial aid administrator (FAA) may use professional judgment to make adjustments on a case-by-case basis, which must be documented, to the cost of attendance (COA)[69] or to the values

---

[66] The campus-based programs—the Federal Supplemental Educational Opportunity Grant (FSEOG) program, the Federal Work-Study program, and the Federal Perkins Loan program—provide federal funding to IHEs for the provision of need-based financial aid to students. For more information on the campus-based aid programs, see CRS Report RL31618, *Campus-Based Student Financial Aid Programs Under the Higher Education Act*, by Joselynn H. Fountain.

[67] The purpose of the Federal Work-Study program is to provide part-time employment to undergraduate, graduate, and professional students in need of earnings to pursue their course of study and to encourage student participation in community service activities.

[68] The EFC is the amount expected to be contributed by the student and the student's family toward postsecondary education expenses for the upcoming award year.

[69] The COA is a measure of a student's educational expenses for the period of enrollment.

of the items used in calculating the EFC to reflect a student's special circumstances in such a way as to make the student eligible for more Title IV assistance.

- Students may maintain their eligibility to receive Title IV federal student aid, despite not making satisfactory academic progress, if the IHE documents that the failure to maintain satisfactory academic progress was due to a disaster.
- Students may be granted a leave of absence during a clock hour or nonterm program, and the requirement that a leave of absence must be requested in writing may be waived.
- The requirement that a student must return a Title IV grant overpayment if the student withdraws from the IHE because of a disaster may be waived under specified conditions.
- Students may continue to receive FWS payments, up to the length of the award period, during periods that they are unable to fulfill their FWS employment obligations as a result of a disaster.
- Borrowers of Direct Loan program loans, Federal Family Education Loan (FFEL) program loans, and Perkins Loans who were in an "in-school" status[70] on the date their attendance was interrupted due to a disaster may remain in that status through the end of the enrollment period in which the disaster occurred.
- Borrowers of DL program loans, FFEL program loans, and Perkins Loans who are in repayment status at the time a disaster occurs and who are unable to make loan payments due to the disaster may be granted forbearance.[71]
- Collection activities on defaulted DL program loans, FFEL program loans, and Perkins Loans may be suspended for three months for borrowers who have been affected by a disaster.

---

[70] During "in-school" status, borrowers need not make payments on the principal and interest of their federal student loans.

[71] Under forbearance, borrowers may temporarily cease making payments on their federal student loans. Forbearance is limited to an initial period of three months for FFEL program loans and Perkins Loans.

- Borrowers of defaulted DL program loans, FFEL program loans, and Perkins Loans who have been affected by a disaster and who are attempting to rehabilitate their loans, consolidate them out of default, or re-establish eligibility for Title IV federal student aid and who fail to make one or more payments while affected by the disaster shall not be considered to have missed any of the required consecutive monthly payments.

Waiver Authority under the Higher Education Relief Opportunities for Students Act (HEROES Act)

In addition to the above-listed waiver authority and regulatory flexibility for Title IV student aid programs, additional authority is provided through the Higher Education Relief Opportunities for Students Act (HEROES Act).[72] The HEROES Act can only be implemented, however, in connection with a war or other military action or a national emergency declared by the President of the United States. The HEROES Act provides the Secretary with authority to waive or modify statutory and regulatory requirements that apply to the HEA Title IV student aid programs in an effort to help affected individuals. There are three categories of affected individuals:

1) Individuals who are serving on active duty or performing qualifying National Guard duty during a war or other military operation or national emergency;
2) Individuals who reside or are employed in an area that is declared a disaster area by any federal, state, or local official in connection with a national emergency; and

[72] The provisions were originally enacted by the Higher Education Relief Opportunities for Students Act of 2001 (2001 HEROES Act; P.L. 107-122; 20 U.S.C. 1070 note) with an expiration date of September 30, 2003. The Higher Education Relief Opportunities for Students Act of 2003 (2003 HEROES Act; P.L. 108-76; 20 U.S.C. 1070 note), provided for waiver authority and regulatory flexibility from FY2003-FY2005; it was extended by P.L. 109-78 to September 30, 2007, and finally made permanent by P.L. 110-93 (20 U.S.C. 1098aa et seq.).

3) Individuals who suffered direct economic hardship as a direct result of a war or other military operation or national emergency.[73]

Provisions for IHEs

Examples of support that may be available under the HEROES Act to IHEs in the aftermath of a war or other military action or national emergency declared by the President of the United States include the following:

- With respect to all categories of affected individuals (and their spouses and dependent children, if applicable), IHEs are excepted from the requirement that financial aid administrators' (FAAs') exercise of professional judgment to make adjustments to the cost of attendance (COA) or to the values of items used in calculating their expected family contribution (EFC) be done on a case-by-case basis.
- With respect to all affected individuals, IHEs are excepted from the requirements to notify affected individuals of a Title IV grant overpayment and the actions that must be taken to resolve the overpayment, to refer a grant overpayment to the Department of Education (ED) under certain conditions, and to deny eligibility for federal student aid to students who do not take action to resolve an overpayment.
- With respect to all affected individuals, IHEs may accept alternative documentation to verify the income and taxes paid by those who are FSA applicants and who are granted an income tax return filing extension by the Internal Revenue Service (IRS).
- With respect to students who withdraw because they are affected individuals in either the 1st or 2nd categories, IHEs that have

[73] For information on the current waivers and modifications issued, see Office of Postsecondary Education, Department of Education, "Federal Student Aid Programs (Student Assistance General Provisions, Federal Perkins Loan Program, Federal Family Education Loan Program, and the Federal Direct Loan Program)," 82 *Federal Register* 45465-45471, September 29, 2017. These currently available waivers and modifications will expire on September 30, 2022.

attempted to contact an affected student (or the parent of the student) regarding the payment of any Title IV credit balance within 14 days of the Title IV credit balance occurring, are considered to have met the 14-day requirement that applies to the payment of Title IV credit balances.

- With respect to the 1st and 2nd categories of affected individuals, IHEs may disburse Title IV funds to a bank account designated to a student or a parent; use Title IV funds to pay charges other than tuition and fees and, as applicable, room and board; or hold Title IV funds on behalf of students on the basis of oral consent, as opposed to with written consent, which would otherwise be required.
- With respect to the 1st and 2nd categories of affected individuals, IHEs are excluded from the requirement to attempt to collect or recover amounts owed from defaulted Perkins Loans for the period during which the borrower is an affected individual, and during a three-month transition period that follows. (Similar exclusions are granted to guaranty agencies with respect to amounts owed on defaulted FFEL program loans, and the Secretary will act similarly with respect to loans held by ED.)

Provisions for Student Financial Aid Recipients

Examples of support that may be available under the HEROES Act to Title IV aid recipients who are unable to meet the specified Title IV statutory or regulatory requirements because of their status as affected individuals include the following:

- Typically, a student's expected family contribution (EFC) is based on financial information from the second preceding tax year. With respect to all affected individuals, an IHE's FAA may exercise professional judgment to consider more recent measures of income. Specifically, the FAA may base an affected individual's EFC on adjusted gross income plus untaxed income and benefits received in the first calendar year of the award year or based on another annual income that more accurately reflects a family's current financial circumstances. As with all cases of professional judgment, the FAA must document the reasons and supporting facts for any adjustment.

- With respect to all affected individuals, the requirement that any unearned grant funds must be returned or repaid may be waived.
- With respect to the 1st and 2nd categories of affected individuals, students who are eligible for a post-withdrawal loan disbursement must be provided at least 45 days, rather than 14 days, to respond and accept the disbursement.
- With respect to the 1st and 2nd categories of affected individuals, the requirement that a leave of absence must be requested in writing may be waived for students who must interrupt their enrollment.
- With respect to students who withdraw because they are affected individuals in either the 1st or 2nd categories, if an IHE has attempted to contact a student regarding the option to have any Title IV credit balance used to reduce the student's Title IV loan debt, the student must be provided up to 45 days to respond to the institution's request.
- Students who are in the 1st or 2nd categories of affected individuals must be permitted 60 days, rather than 14 days, to request the cancellation of all or a portion of a loan or TEACH Grant.
- Students who are in the 1st or 2nd categories of affected individuals may continue to receive Title IV funds despite not making satisfactory progress in coursework.[74]
- For borrowers of DL program loans, FFEL program loans, and Perkins Loans who are in the 1st or 2nd categories of affected individuals, the initial grace period[75] excludes any period, not to

[74] To be eligible for federal student aid, students must make satisfactory academic progress. That is, they must show satisfactory and timely completion of coursework.

[75] The grace period is the period of time after a student graduates, leaves school, or drops below half-time enrollment before he or she must begin repaying his or her federal student loan. Grace periods vary by loan type, but typically are 6-9 months in length.

exceed three years, during which a borrower is an affected individual.

- Borrowers of DL program loans, FFEL program loans, and Perkins Loans who were in an "in-school" status but who left school because they became a 1st or 2nd category affected individual may retain their in-school status for up to three years. During this period, the Secretary will pay any interest that accrues on a Subsidized Stafford Loan.
- Borrowers of DL program loans, FFEL program loans, and Perkins Loans who were in an "in-school" deferment or a graduate fellowship deferment but who left school because they became a 1st or 2nd category affected individual may retain their deferment for a period of up to three years during which they are affected individuals. During this period, the Secretary is to pay any interest that accrues on a Subsidized Stafford Loan.
- For borrowers of Perkins Loans who are in the 1st or 2nd categories of affected individuals, any forbearance granted on the basis of the borrower's status as an affected student is excluded from the usual three-year limit on forbearance. Also, for these categories of affected individuals, borrowers of Perkins Loans may be granted forbearance based on an oral request and without written documentation for a one-year period and an additional three-month transition period.
- For borrowers of FFEL program loans who are in the 1st or 2nd categories of affected individuals, borrowers may be granted forbearance based on an oral request and without written documentation for a one-year period and an additional three-month transition period.
- For borrowers of FFEL program loans, DL program loans, and Perkins Loans that may qualify for cancellation on the basis of continuous or uninterrupted qualifying service, such service will not be considered interrupted by any period during which the borrower is an affected individual in the 1st or 2nd categories or during a three-month transition period.

- With respect to the 1st and 2nd categories of affected individuals, for borrowers of defaulted FFEL or DL program loans seeking to rehabilitate their loans by making 9 on-time payments in 10 consecutive months and borrowers of defaulted Perkins Loans seeking to rehabilitate their loans by making 9 consecutive on-time payments, any payments missed during the period when the borrowers are affected individuals or during the three-month transition period that follows shall not be considered an interruption of the series of payments required for loan rehabilitation.
- With respect to the 1st and 2nd categories of affected individuals, for borrowers of defaulted FFEL program loans, DL program loans, or Perkins Loans seeking to reestablish eligibility for Title IV federal student aid by making six consecutive on-time payments, any payments missed during the period when the borrowers are affected individuals or during the three–month transition period that follows shall not be considered an interruption of the series of payments required for purposes of reestablishing Title IV eligibility.
- With respect to the 1st and 2nd categories of affected individuals, for borrowers of defaulted FFEL or DL program loans seeking to consolidate loans out of default, any payments missed during the period when the borrowers are affected individuals or during the three–month transition period that follows shall not be considered an interruption of the series of payments required for purposes of reestablishing Title IV aid eligibility.
- With respect to the 1st and 2nd categories of affected individuals, for borrowers who are repaying their FFEL or DL program loans according to the Income-Based Repayment (IBR), Income-Contingent Repayment (ICR), Pay As You Earn (PAYE) or Revised Pay As You Earn (REPAYE) plan and who because of their status as affected individuals are unable to provide information normally required annually to document their income and family size may maintain their current payment amount for a

period of up to three years, including a three-month transition period immediately following. This flexibility is made in lieu of having their payment amount adjusted to be based on a standard 10-year repayment plan or an alternative repayment plan, as applicable.

- With respect to students who are dependents of affected individuals in the 1st category, the requirement to submit a statement signed by a parent verifying the number of family members in the household and the number enrolled in postsecondary institutions is waived if no responsible parent is available to provide a signature.
- With respect to students who are dependents of affected individuals in the 1st category, the requirement for at least one parental signature on the Free Application for Federal Student Aid (FAFSA), the Student Aid Report (SAR), and the Institutional Student Information Record (ISIR) is waived if no responsible parent is available to provide a signature. A student's FAA or high school counselor may sign on behalf of the parent, so long as the applicant provides adequate documentation regarding the parent's inability to provide a signature due to his or her status as an affected individual.

HEA Maintenance of Effort Requirements for the College Access Challenge Grant Program

The College Access Challenge Grant Program (CACG; HEA Title VII, Part E) fostered partnerships between federal, state, and local governments and philanthropic organizations through matching formula grants that were intended to increase the number of low-income students who are prepared to enter and succeed in postsecondary education. To maintain eligibility for the CACG, a state was required to meet maintenance of effort requirements. According to the MOE requirements, a state was required to annually provide

- for public IHEs in the state, at least the average amount provided for noncapital and nondirect research and development expenses or costs in the five preceding academic years and
- for private IHEs in the state, at least the average amount provided for postsecondary student financial aid to such IHEs in the five preceding academic years.

The Secretary was authorized to waive the MOE requirements for a state in the event of "exceptional or uncontrollable circumstances, such as a natural disaster or a precipitous and unforeseen decline in the financial resources of a state or state educational agency."[76] The authority to award new CACGs expired at the end of FY2014.

Titles III and V Endowment Funds

Title III, Parts A, B, and F and Title V, Part A of the HEA authorize several programs that provide institutional aid to IHEs that serve a high proportion of low-income students and that have low educational and general expenditures per students in comparison to similar institutions.[77] Recipients of awards from the programs, except the Native American-Serving Nontribal Institutions programs, may use a portion of their awards to establish or increase an endowment fund. In addition, Title III, Part C of the HEA authorizes the Endowment Challenge Grants program, which allows recipients to use their awards to establish or increase an endowment fund. While the endowment fund corpus for each of the programs must be maintained for 20 years, 50% of the accumulated endowment fund income may be used for eligible institutional costs. As a consequence of a financial emergency, a life-threatening situation occasioned by a natural disaster or arson, or unusual occurrence or exigent circumstance, the Secretary may permit an IHE to spend more than 50% of the endowment fund income.[78]

---

[76] HEA §137.

[77] For additional information on the Title III/V programs, see CRS Report R43237, *Programs for Minority-Serving Institutions Under the Higher Education Act*, by Alexandra Hegji.

[78] Congress last appropriated funds for the Endowment Challenge Grant Program in FY1995. However, several institutions continue to participate in the program, as they have not yet met the 20-year maintenance date.

Education Disaster and Emergency Relief Loan Program

Section 824 of the HEA authorizes the Secretary of Education, in consultation with the Secretary of Homeland Security, to establish an Education Disaster and Emergency Relief Loan Program for IHEs impacted by a major disaster or emergency declared by the President. Funds may be used for construction, replacement, renovation, and operations costs resulting from the disaster or emergency. The program has never been funded.

Temporary HEA Provisions Enacted in Response to Disasters

In response to several natural disasters over the years, previous Congresses enacted numerous temporary programs and provisions (e.g., waiver authorities for specific programs) to support education. A brief summary of the various programs and provisions created in response to previous disasters (e.g., Hurricanes Katrina and Rita) is included in the Appendix. For example, additional funds were appropriated to IHEs to support recovery efforts and to provide grants to students. In addition, Congress authorized the Secretary to modify or waive certain program requirements to ensure that IHEs affected by a natural disaster did not lose grant funding as a result of the disaster, and the Secretary was given special authority to ensure historically black colleges and universities were able to gain access to capital financing to support their recovery. Much of the additional funding was provided up through FY2010 and many of the temporary provisions (expired by FY2014). Most recently, legislation was enacted that authorizes temporary education-related supplemental funds and flexibilities in response to several 2017 natural disasters, including Hurricanes Harvey, Irma, and Maria and the 2017 California wildfires.

The Hurricanes Harvey, Irma, and Maria Education Relief Act of 2017

In the 115th Congress, in response to multiple 2017 natural disasters, the Hurricanes Harvey, Irma, and Maria Education Relief Act of 2017 was enacted.[79] As it relates to postsecondary education, the act makes several modifications to the Federal Work Study (FWS) and Federal Supplemental Educational and Opportunity Grant (FSEOG) programs:

- With respect to funds made available for award years 2016-2017 and 2017-2018, the act requires the Secretary to waive nonfederal share requirements of FWS and FSEOG programs[80] for IHEs located in affected areas[81] and authorizes the Secretary to waive such requirements for IHEs not located in an affected area that have enrolled or accepted for enrollment any affected students,[82] after considering such IHE's student population and existing resources.
- The act requires the Secretary to reallocate unexpended federal FWS or FSEOG funds that were returned to the Secretary by an IHE for award year 2016-2017[83] to eligible IHEs (with priority

[79] P.L. 115-64.

[80] Under each of the programs, IHEs must provide nonfederal matching funds, typically equal to 25% of program funds. For additional information on the FWS and FSEOG programs, see CRS Report RL31618, *Campus-Based Student Financial Aid Programs Under the Higher Education Act*, by Joselynn H. Fountain.

[81] An affected area is defined as "an area for which the President declared a major disaster or an emergency under section 401 or 501, respectively, of the Robert T. Stafford Disaster Relief and Emergency Assistance Act ... as a result of Hurricane Harvey, Hurricane Irma, Hurricane Maria, Tropical Storm Harvey, Tropical Storm Irma, or Tropical Storm Maria." P.L. 115-64.

[82] An affected student is an individual who has applied for or received HEA Title IV student aid and who (1) was enrolled or accepted for enrollment at an IHE located in an affected area on August 25, 2017; (2) is a dependent student who was enrolled or accepted for enrollment at an IHE not located in an affected area on August 25, 2017, but whose parents resided or were employed in an affected area on August 25, 2017; or (3) "suffered direct economic hardship" as a result of Hurricanes Harvey, Irma, or Maria or Tropical Storms Harvey, Irma, or Maria, as determined by the Secretary; P.L. 115-64.

[83] IHEs that participate in the FWS and FSEOG programs may return to ED unexpended program funds for any fiscal year, and the Secretary generally is required to reallocate such funds to other eligible institutions participating in the programs. Reallocation methodologies vary between the programs. See 34 C.F.R. §673.4. ED recently announced it has already reallocated FSEOG funds to IHEs that have been directly affected by the 2017 hurricanes

given to IHEs located in affected areas) and to waive the FWS and FSEOG allocation reduction[84] for award year 2018-2019 for eligible IHEs.[85]

- The act authorizes the Secretary to waive or modify any other statutory or regulatory provision relating to the reallocation of excess FWS or FSEOG funds to ensure that assistance is received by eligible IHEs.
- The act makes available for obligation by the Secretary until September 20, 3018, FWS and FSEOG funds for which the period of availability otherwise expired on September 30, 2017.
- The act authorizes the Secretary to recall any FWS or FSEOG funds allocated to an IHE in award year 2016-2017 that are not returned to ED as excess allocations but that otherwise would have lapsed on September 30, 2017, and to reallocate those funds to eligible IHEs.

All of these provisions expire on September 30, 2018.

The Further Additional Supplemental Appropriations for Disaster Relief Act, 2018

Also in the 115th Congress, in response to Hurricanes Harvey, Irma, and Maria and wildfire incidents in California in 2017, the Further Additional Supplemental Appropriations for Disaster Relief Act, 2018 (Division B, Subdivision 1 of the Bipartisan Budget Act of 2018, P.L. 115-123) was enacted. As it relates to postsecondary education, the act includes FY2018 supplemental appropriations for disaster relief for "covered disasters or emergencies" (i.e., Hurricanes Harvey, Irma, Maria, or

---

and is reviewing the reallocation process for FWS funds. U.S. Department of Education, Updates Information Related to the 2017-2018 Supplemental Campus-Based Funds Due to 2017 Hurricanes, September 29, 2017, https://ifap.ed.gov/eannouncements/092917UpdateInfoRelate20172018SupplementalCampusBasedFunds2017Hurricanes.html.

[84] Typically, if an IHE returns more than 10% of its FWS or FSEOG allocation in any fiscal year, its allocation for the next fiscal year is reduced by the amount returned.

[85] An eligible IHE is an IHE that participates in the relevant program (i.e., FWS and/or FSEOG) and is located in an affected area or has enrolled or accepted for enrollment affected students in award year 2017-2018.

wildfires in 2017 for which a major disaster or emergency has been declared under section 401 or 501 of the Robert T. Stafford Disasters Relief and Emergency Assistance Act). Specifically, the supplemental appropriations for postsecondary education include the following:

- $100 million for the FSEOG program, the FWS program, and the Fund for the Improvement of Postsecondary Education. Funds shall be for IHEs located in areas affected by a covered disaster or emergency and students enrolled in such institutions. Matching requirements under FSEOG and FWS are waived for IHEs receiving such funds. Funds may be used for a variety of expenses including student financial assistance, staff salaries, student supplies and equipment, or any other purpose authorized under the HEA.
- Up to $75 million for IHEs to help defray the unexpected expenses associated with enrolling students displaced from IHEs at which operations have been disrupted by a covered disaster or emergency.

In addition, the act authorizes several flexibilities in the administration of several HEA programs, including the following:

- Authorization for the Secretary to waive, modify, or provide extensions for any statutory or regulatory provisions applicable to the HEA Title IV programs or student eligibility or institutional eligibility provisions in the HEA for affected individuals, students, and IHEs in covered disaster or emergency areas, if the Secretary deems such waiver, modification, or extension necessary in connection with a covered disaster or emergency.[86]
- Authorization for the Secretary to modify required and allowable uses of funds under the various minority-serving grant programs

[86] Authorized flexibilities mirror those authorized following the 2005 hurricane season under the Higher Education Hurricane Relief Act of 2005 (P.L. 109-148).

authorized under HEA Title III, Parts A and B, and under the TRIO and GEAR-UP programs.

- Authorization for the Secretary to forgive any outstanding balance owed to ED under the HBCU Hurricane Supplemental Loan program established under P.L. 109-234 and in response to the 2005 hurricane season.[87]

Veterans Educational Assistance

In general, the GI Bills and the Vocational Rehabilitation and Employment program provide individuals with a certain entitlement to educational benefits. For example, under the Post-9/11 GI Bill, individuals are entitled to 36 months of benefits.

Statutory provisions allow the VA to continue paying benefits during interval periods when a school is temporarily closed due to an emergency situation, as long as the aggregate periods do not exceed four weeks during a 12-month period. In the event of mitigating circumstances (circumstances beyond the individual's control that prevent continuous pursuit of a program of education), the VA may pay benefits to an individual for a course from which the individual withdraws. Under normal circumstances, interval payments and payments for withdrawn courses reduce an individual's entitlement. However, the individual's entitlement will not be reduced if an individual requests interruption of benefits for any break when a school was closed during a certified period of enrollment and payments were continued due to an emergency situation.

The Survivors' and Dependents' Educational Assistance Program (DEA; Title 38 U.S.C., Chapter 35) pays educational benefits to children

---

[87] Under this provision, four HBCUs—Dillard University, Southern University at New Orleans, Xavier University, and Tougaloo College—are eligible to have their loans forgiven. It is reported that these four schools borrowed a total of $361 million in loans capped at a 1% interest rate to be repaid over 30 years. They have repaid approximately $12.4 million of the loans since borrowing them. Melissa Korn, "Loan Program for Black Colleges Struggles with Oversight, Repayment; Some of the $2 billion in loans ends up more a burden than a help to the schools," *Wall Street Journal*, July 17, 2017.

after they achieve a high school diploma or its equivalent, or after they reach 18 years of age, but before they reach 26 years of age. In certain situations, such as an enrollment suspension due to conditions determined by the VA to have been beyond the individual's control, benefits may be extended beyond an eligible person's 26th birthday, but generally not past the 31st birthday.

OTHER PROGRAMS ADMINISTERED BY THE DEPARTMENT OF EDUCATION

The Department of Education administers various other programs related to education that are not necessarily classified as serving elementary and secondary education or postsecondary education purposes. Two such programs are state grants authorized under the Adult Education and Family Literacy Act (AEFLA), which support education for adults at the secondary level and below, as well as English language training; and Vocational Rehabilitation State Grants authorized under the Rehabilitation Act of 1973, which support the provision of vocational rehabilitation (VR) services to individuals with disabilities.

This section provides an overview of the authorized statutory and regulatory authorities that may be useful in supporting education-disaster recovery within these programs. Generally, current law permits disaster-affected grantees under these programs to receive a waiver for each program's maintenance of effort requirements and certain timely reporting requirements but does not permit waivers for required state matching funds.

Waiver Authority and Flexibilities for AEFLA State Grants

The Adult Education and Family Literacy Act (AEFLA) provides grants to states to support education for out-of-school adults at the secondary level and below, as well as English language training.

AEFLA requires states to match a portion of the federal grant and has a maintenance of effort requirement (MOE) that reduces a state's federal grant if the state reduces its nonfederal support. Statute authorizes the Secretary to waive the MOE requirement if "a waiver would be equitable due to exceptional or uncontrollable circumstances, such as a natural disaster or an unforeseen and precipitous decline in the financial resources."[88] Statute specifies that if the Secretary grants a waiver for a given fiscal year, the level of required effort of meet the MOE requirement shall not be reduced in subsequent years because of the waiver. AEFLA does not grant ED the authority to waive states' nonfederal match.

In addition, AEFLA state grantees are required to submit annual performance reports as a condition of receiving funding. AEFLA performance reports are included with reports for other federally supported workforce grant programs. Regulations specify that a state will not be subject to financial sanctions if the state fails to submit a complete and timely report due to exceptional circumstances outside the state's control, including a disaster.[89]

Waiver Authority and Flexibilities for Vocational Rehabilitation State Grants

Under the authorization of the Rehabilitation Act of 1973, as amended (the Rehabilitation Act), ED provides grants to state agencies to provide vocational rehabilitation (VR) services to individuals with disabilities. States may use grant funds to assess clients and help eligible clients develop and carry out an individualized plan for employment.

The Rehabilitation Act requires states to match a portion of the federal grant and has a MOE requirement that reduces a state's federal grant if the state reduces its nonfederal contribution. Statute authorizes the Secretary to waive the MOE requirement if a waiver "is necessary to permit the State to respond to exceptional or uncontrollable circumstance, such as a major

---

[88] 29 U.S.C. 3331(b)(4).

[89] See 34 C.F.R. 361.185. This section of regulations also applies to other partner programs.

natural disaster or serious economic downtown."[90] The Rehabilitation Act does not grant ED the authority to waive states' nonfederal match.[91]

In addition, VR state grantees are required to submit annual performance reports as a condition of receiving funding. VR performance reports are included with reports for other federally supported workforce grant programs. Regulations specify that a state will not be subject to financial sanctions if the state fails to submit a complete and timely report due to exceptional circumstances outside the state's control, including a disaster.[92]

APPENDIX. EDUCATION-RELATED FEDERAL DISASTER APPROPRIATIONS AND TEMPORARY PROVISIONS ADMINISTERED BY THE U.S. DEPARTMENT OF EDUCATION FOLLOWING VARIOUS NATURAL DISASTERS

Between 2005 and 2010, Congress appropriated nearly $2 billion for ED to provide support to local educational agencies (LEAs), schools, and institutions of higher education (IHEs) that were affected by Hurricanes Katrina, Rita, or Gustav and other natural disasters. Funding provided for elementary and secondary education and for higher education is discussed below. Table A-1 details how much of this funding was allocated specifically to Alabama, Florida, Louisiana, Mississippi, Texas, and all other states combined, by program. In addition, temporary provisions and flexibilities provided for elementary and secondary education, higher education, and the Vocational Rehabilitation State Grant program are described below.

---

[90] Quoted text from 34 C.F.R. 361.62; MOE requirement and waiver authority are in statute at 29 U.S.C. 731(a).

[91] In some circumstances, Congress has passed legislation to temporarily waive the nonfederal match for states affected by a natural disaster. See discussion of P.L. 109-82 in the Appendix.

[92] See 34 C.F.R. 361.185.

Elementary and Secondary Education[93]

Following the Gulf Coast hurricanes, funding to support elementary and secondary schools affected by Hurricane Katrina or Hurricane Rita was provided through three public laws: P.L. 109-148 ($1.4 billion), P.L. 109-234 ($235 million), and P.L. 110-28 ($30 million).

- P.L. 109-148 created two new programs: (1) Immediate Aid to Restart School Operations ($750 million) and (2) Temporary Emergency Impact Aid for Displaced Students ($645 million), specifically designed to address needs resulting from the hurricanes and to provide support to LEAs through an existing federal education program administered by ED.[94] It also added $5 million to the McKinney-Vento Homeless Assistance Act to serve homeless children and youth who had been displaced by the Gulf Coast hurricanes.
- P.L. 109-234 provided additional funding of $235 million for the Temporary Emergency Impact Aid for Displaced Students enacted under P.L. 109-148.
- P.L. 110-28 appropriated $30 million for elementary and secondary schools affected by the hurricanes through the Hurricane Educator Assistance program to assist in recruiting, retaining, and compensating staff in those schools.

---

93 While not provided through education-related legislation enacted in response to a disaster, Louisiana also received $20.9 million through the Charter School Program authorized under Title V-B-1 of the Elementary and Secondary Education Act (prior to enactment of the ESSA) specifically to help reopen charter schools damaged by Hurricane Katrina and Hurricane Rita, help create 10 new charter schools and expand existing charter schools to accommodate displaced students. (For more information, see U.S. Department of Education, "Louisiana Awarded $20.9 Million No Child Left Behind Grant to Assist Damaged Charter Schools, Create New Charter Schools," press release, September 30, 2005, http://www2.ed.gov/news/pressreleases/2005/09/09302005.html).

94 In addition to funding, P.L. 109-148 provided general waiver authority for the Secretary of Education related to maintenance of effort (MOE) requirements; the use of federal funds to supplement, not supplant nonfederal funds; and matching contributions for programs administered by the Secretary. It also modified hold harmless provisions for the Elementary and Secondary Education Act (ESEA) Title I-A Grants to Local Educational Agencies program and modified highly qualified teacher provisions contained in ESEA Title I-A.

Congress then appropriated an additional $15 million through P.L. 110-329 to provide support to LEAs whose enrollment of homeless students increased as a result of hurricanes, floods, or other natural disaster during 2008. Most recently, Congress appropriated $12 million through P.L. 111- 117 for the Gulf Coast Recovery Initiative to improve education in areas affected by Hurricanes Katrina, Rita, or Gustav. A brief description of each of these programs and the amount of funding each received is presented below. Table A-1 details the funding amounts various states received under each of the programs.

Immediate Aid to Restart School Operations

The Immediate Aid to Restart School Operations provided support for LEAs and nonpublic schools in Louisiana, Mississippi, Alabama, and Texas to restart school operations, reopen schools, and re-enroll students. P.L. 109-148 provided $750 million for this program. This program is no longer authorized.

Temporary Emergency Impact Aid for Displaced Students

The Temporary Emergency Impact Aid for Displaced Students program provided federal funding to assist schools in enrolling students who had been displaced by the Gulf Coast hurricanes. Funds were made available to LEAs and schools based on the number of displaced students that enrolled, irrespective of whether the school in which parents chose to enroll their child was a public or nonpublic school. P.L. 109-148 appropriated $645 million for this program. Subsequently, P.L. 109-234 appropriated an additional $235 million for this program, bringing the total program appropriation to $880 million.[95] Portions of the funds appropriated were provided to 49 states[96] and the District of Columbia based on the number of displaced students each enrolled. Louisiana, Texas,

---

[95] Of the total appropriation for Temporary Emergency Impact Aid for Displaced Students, only $878 million was distributed, as the remaining funds were not needed by states under this program.

[96] Hawaii did not receive any funds through this program.

and Mississippi received the largest proportion of funds. This program is no longer authorized.

McKinney-Vento Homeless Assistance Act

The McKinney-Vento Homeless Assistance Act (P.L. 100-77; 42 U.S.C. 11433 et seq.) provides funding to states to ensure that homeless children and youth are provided equal access to a free, appropriate public education in the same manner as provided other children and youth through the Education for Homeless Children and Youths program.[97] P.L. 109-148 appropriated $5 million for this program for LEAs serving homeless children and youth who had been displaced by Hurricanes Katrina or Rita. Eight states received funding under this program, with the largest grants provided to Texas and Louisiana.[98] While the McKinney-Vento Homeless Assistance Act continues to provide funding related to the education of homeless students, the provisions enacted specifically in response to the Gulf Coast hurricanes are no longer authorized.

Hurricane Educator Assistance Program

The Hurricane Educator Assistance Program made federal funding available to Louisiana, Mississippi, and Alabama to use for recruiting, retaining, and compensating school staff who committed to work for at least three years in public elementary and secondary schools affected by Hurricanes Katrina or Rita. States were required to apply to receive funds, and the funds were allocated based on the number of public elementary and secondary schools that were closed for 19 days or more from August 29, 2005, through December 31, 2005. P.L. 110-28 provided $30 million for these purposes. Funds were provided to Louisiana and Mississippi only. This program is no longer authorized.

---

[97] For a description of existing federal programs enacted by the McKinney-Vento Homeless Assistance Act, see CRS Report RL30442, *Homelessness: Targeted Federal Programs*, coordinated by Libby Perl.

[98] The eight states that received funds were Alabama, Arkansas, Florida, Georgia, Louisiana, Mississippi, Tennessee, and Texas.

Homeless Education Disaster Assistance

P.L. 110-329 provided $15 million to LEAs whose enrollment of homeless students increased as a result of hurricanes, floods, or other natural disasters that occurred during 2008 and for which the President declared a major disaster under Title IV of the Stafford Act. ED was required to distribute the funds through the McKinney-Vento Homeless Assistance Act based on demonstrated need. These funds provided assistance to LEAs in Gulf Coast states affected by Hurricanes Gustav and Ike, as well as LEAs affected by natural disasters in other parts of the nation, such as flooding in the Midwest. Of the $15 million provided, $13.5 million was provided to Florida, Louisiana, and Texas.[99] This program is no longer authorized.

Gulf Coast Recovery Initiative

P.L. 111-117 provided $12 million for competitive awards to LEAs located in counties in Louisiana, Mississippi, and Texas that were designated by FEMA as counties eligible for individual assistance as a result of damage caused by Hurricanes Katrina, Rita, or Gustav. The funds had to be used to improve education in areas affected by these hurricanes and had to be used for activities such as replacing instructional materials and equipment; paying teacher incentives; modernizing, renovating, or repairing school buildings; supporting charter school expansion; and supporting extended learning time activities. The majority of the funds were provided to LEAs in Louisiana. This program is no longer authorized.

---

[99] While data were not available on the specific disasters experienced by the LEAs that received funding, data were available on the specific types of disasters for which IHEs received funds under the Higher Education Disaster Relief program (P.L. 110-329), which also provided aid in response to natural disasters that occurred in 2008. According to these data, all IHEs in Louisiana that received funds were affected by Hurricanes Gustav or Ike. Most IHEs in Texas that received funds were affected by Hurricane Ike. A few IHEs in Texas were affected by Hurricane Dolly, accounting for a relatively small portion of the funds allocated to IHEs in Texas. IHEs in Florida that received funding were affected by Tropical Storm Fay. LEAs in Iowa and Illinois received the remaining funds available to LEAs.

Temporary Flexibilities under the ESEA

P.L. 109-148 provided the Secretary with the authority to grant LEAs certain temporary flexibilities and waivers from fiscal requirements under the ESEA. These included the temporary authority to waive or modify any requirements related to maintenance of effort (MOE)[100]; the use of federal funds to supplement, not supplant, nonfederal funds; and any nonfederal share or capital contribution required to match federal funds provided under programs administered by the Secretary. The waivers could be provided at the Secretary's discretion to ease the fiscal burdens of entities in an affected state in which a major disaster has been declared in accordance with Section 401 of the Stafford Act. It is important to note that the waivers of MOE did not reduce the level of effort required for the subsequent year. It should also be noted that this additional authority to waive fiscal accountability requirements did not permit the Secretary to waive or modify any provision under IDEA.

In response to multiple disasters, Congress has required the Secretary to hold affected LEAs harmless at 100% of their prior-year grant amount under Title I-A of the ESEA.[101] It appears that these provisions may have been enacted as a means of preventing affected LEAs from experiencing a decrease in funding due to the displacement of students.[102]

---

[100] As previously noted, ESEA §8521(c) provides the Secretary with the authority to waive MOE requirements for LEAs under specific circumstances.

[101] For more information, see P.L. 109-148 and P.L. 110-366.

[102] LEA grant allocations under Title I-A of the ESEA are largely based on counts of children aged 5-17 living in poverty. These counts are estimated by the U.S. Census Bureau annually through the Small Area Income Poverty Estimates (SAIPE). However, there is a two-year delay between the year for which child counts are estimated and the year for which those estimates are used in the allocation of funds. For example, data used to make grants for the 2016- 2017 school year (FY2016) are based on data from the 2014 calendar year. Thus, holding LEAs harmless in the short term does not actually reflect student displacement as a result of the disaster. Rather, the effects of student displacement would not affect Title I-A grants until about two years after the disaster has occurred. The hold harmless provisions enacted in response to prior disasters have not taken this into account. At the same time, the implementation of these hold harmless provisions may have limited any increases in the allocations available to LEAs that gained formula children during the calendar year for which counts were used.

Temporary Flexibilities under the IDEA

P.L. 109-148 also provided the Secretary with the temporary authority to extend certain administrative deadlines under the IDEA, such as the deadline for submission of an annual report on the progress of the state and of children with disabilities in the state for the 2005-2006 academic year.[103] In addition, the Secretary was authorized to extend the deadline for the initial evaluation of children to determine their eligibility for services under IDEA. In providing the Secretary with the authority to extend these deadlines, Congress also included statutory language that clarified that this new authority should not be interpreted as granting the Secretary waiver authority over key IDEA provisions, such as the right to a free appropriate public education or procedural safeguards granted under the act.

Higher Education

Appropriations to support IHEs following the Gulf Coast hurricanes of 2005 (i.e., Hurricanes Katrina and Rita) were provided in 2005 through P.L. 109-148 ($200 million), in 2006 through P.L. 109-234 ($50 million), and in 2007 through P.L. 110-28 ($30 million). P.L. 110-329 subsequently provided another $15 million for IHEs in areas affected by hurricanes, floods, or other natural disasters in 2008. Table A-1 details the amount of funding allocated to states under these provisions.

Hurricane Education Recovery

Of the $200 million provided under P.L. 109-148 for higher education, $95 million was specifically appropriated for the Louisiana Board of Regents, and $95 million was specifically appropriated for the Mississippi Institute of Higher Learning for hurricane recovery.[104] Subsequently, P.L. 109-234 and P.L. 110-28 provided additional funds for hurricane recovery under the Fund for the Improvement of Postsecondary Education (FIPSE), authorized by Title VII of the Higher Education Act (HEA), to assist IHEs adversely affected by the 2005 Gulf Coast hurricanes. Under both laws,

---

[103] P.L. 109-148 §301.

[104] Funds could be spent for a variety of purposes, such as faculty and staff salaries, equipment and instruments, and student financial assistance.

funds were provided to help defray the expenses incurred by IHEs that were forced to close, relocate, or reduce their activities due to hurricane damage. Under P.L. 110-28, IHEs also were permitted to use funds to make grants to students enrolled at such institutions on or after July 1, 2006. A total of $80 million was provided for IHEs affected by Hurricane Katrina or Rita under FIPSE for hurricane recovery. The majority of funds appropriated for hurricane education recovery were provided to Mississippi and Louisiana. These activities are no longer authorized.

Funds to Assist IHEs Enrolling Displaced Students

Under P.L. 109-148, $10 million was appropriated to assist IHEs with unanticipated costs associated with the enrollment of students displaced as a result of Hurricane Katrina or Rita. Overall, 99 IHEs in 24 states and the District of Columbia received funds related to the enrollment of displaced higher education students.[105] Louisiana and Texas received the largest state grants. This program is no longer authorized.

Higher Education Disaster Relief[106]

P.L. 110-329 provided an additional $15 million[107] for IHEs that were located in an area affected by hurricanes, floods, and other natural disasters that occurred during 2008 and for which the President declared a major disaster under Title IV of the Stafford Act. Funds provided through the Higher Education Disaster Relief program could be used to defray the expenses incurred by IHEs in affected areas that were forced to close or relocate or whose operations were adversely affected by the natural disaster and to provide grants to students who attended such IHEs for academic years beginning on or after July 1, 2008. The majority of these

---

[105] The 24 states in which IHEs received funds included Alabama, Arizona, Arkansas, California, Colorado, Florida, Georgia, Illinois, Iowa, Kentucky, Louisiana, Maryland, Massachusetts, Michigan, Minnesota, Mississippi, Missouri, New Jersey, New York, Ohio, Tennessee, Texas, Utah, and Virginia.

[106] None of these funds were provided in response to the Gulf Coast hurricanes of 2005.

[107] Based on data available from ED, $15.028 million was provided to IHEs. For more information, see http://www2.ed.gov/programs/disaster-relief/hedr-abstracts2009.pdf.

funds were provided to Louisiana and Texas for hurricane-related education disaster assistance.[108] This program is no longer authorized.

Modified Uses of Funds under Various Grant Programs

Under P.L. 109-148, upon request by an affected institution or other grantee located in an area affected by a 2005 Gulf hurricane disaster (i.e., Hurricane Katrina or Rita), the Secretary was given the authority to modify the required and allowable uses of funds under the TRIO programs (HEA Title IV, Part A, Subpart 2, Chapter 1),[109] Gaining Early Awareness and Readiness for Undergraduate Programs (GEAR UP; HEA Title IV, Part A, Subpart 2, Chapter 2), programs authorized under Strengthening Institutions (HEA Title III, Part A) or Strengthening Historically Black Colleges and Universities (HEA Title III, Part B), and any other competitive grant program. The Secretary was not authorized to permit any new construction, renovation, or improvement of classrooms, libraries, laboratories, or other instructional facilities that was not previously authorized under the institution's grant award under part A or B of Title III.

Modified Requirements under the Title III-A and Title III-B Programs

Congress authorized the Secretary to waive certain requirements under the Title III, Part A and Title III, Part B programs related to institutional eligibility, data submissions, and funding allocations.[110] The waivers were intended to ensure that IHEs affected by either Hurricane Katrina or Rita did not lose funding as a result of the disasters. The waivers were required for FY2009- FY2011 and permitted for FY2012 and FY2013.

---

[108] IHEs in Arkansas, Colorado, Iowa, Illinois, Indiana, and Kentucky also received funds under this program.

[109] For more information on the TRIO programs, see CRS Report R42724, *The TRIO Programs: A Primer*, by Cassandria Dortch.

[110] The waivers (Section 392(c) of the HEA) were enacted by Section 318 of the Higher Education Opportunity Act (HEOA; P.L. 110-315).

Modification of Teacher Quality Enhancement Grants for States and Partnerships

At the request of the grantee, the Secretary may modify the requirements for Teacher Quality Enhancement Grants for States and Partnerships authorized under HEA Title II, Part A.[111] Modifications could be made to aid states and LEAs in recruiting and retaining highly qualified teachers in an LEA in an area affected by a Hurricane Katrina or Rita, and to aid IHEs, located in such an area, in recruiting and retaining faculty needed to prepare teachers and provide professional development.

Additional Capital Financing for Historically Black Colleges and Universities

Congress helped historically Black colleges and universities (HBCUs) affected by the Gulf Coast hurricanes gain access to low-cost capital financing to assist their recovery efforts through the Historically Black College and University (HBCU) Capital Financing program (Title III-D of the HEA). P.L. 109-234 authorized the Secretary to waive or modify statutory or regulatory provisions to ensure that the calculation of financing for an HBCU reflected changes in the financial condition of the institution as a result of the 2005 Gulf Coast hurricanes and to ensure that those affected institutions that were not receiving assistance under the program before the Gulf hurricane disasters were eligible to apply. The law also relaxed several program financing requirements.[112] The Secretary's authority to waive or modify statutory or regulatory requirements of this program is no longer authorized. In 2018, the Further Additional Supplemental Appropriations for Disaster Relief Act, 2018 (Division B, Subdivision 1 of the Bipartisan Budget Act of 2018, P.L. 115-123) authorized the Secretary to forgive any outstanding loan balance owed to ED by institutions that secured funding under this program in the wake of the Gulf Cost Hurricanes.

---

[111] This authority was enacted by the FY2006 Defense Appropriations (P.L. 109-148).

[112] For instance, HBCUs receiving loans under this waiver authority were granted deferments of up to three years. During this time, such HBCUs were not required to make installment payments on loan principal, and interest did not accrue.

Table A-1. Funding Administered by the Department of Education Provided in Response to Natural Disasters: 2005-2010
Dollars in thousands

Department of Education	Alabama	Florida	Louisiana	Mississippi	Texas	All Other States	Total
Elementary and Secondary Education							
Immediate Aid to Restart School Operations	3,750	–	445,604	222,493	78,153	–	750,000
Emergency Impact Aid for Displaced Students[a]	36,605	27,214	291,717	100,787	250,890	170,931	878,144
McKinney-Vento Homeless Education Assistance Program	247	196	1,564	687	1,687	619	5,000
Hurricane Educator Assistance Program	–	–	22,593	7,407	–	–	30,000
Homeless Education Disaster Assistance	–	91	1,171	–	12,256	1,483	15,000
Gulf Coast Recovery Initiative	–	–	8,624	2,638	739	–	12,000
Subtotal for elementary and secondary education	40,602	27,501	771,273	334,012	343,724	173,033	1,690,144
Higher Education							
Hurricane Education Recovery	301	1,507	145,663	117,878	4,651	–	270,000

Table A-1. (Continued)

Department of Education	Alabama	Florida	Louisiana	Mississippi	Texas	All Other States	Total
Higher Education							
Funds to Assist Institutions of Higher Education Enrolling Displaced Students	357	34	5,748	327	1,750	1,783	10,000
Higher Education Disaster Relief Program	–	25	3,524	–	8,354	3,125	15,028
Subtotal post-secondary education	658	1,566	154,935	118,206	14,756	4,908	295,028
Total	41,261	29,067	926,208	452,217	358,480	177,940	1,985,172

Source: Table prepared by CRS based on published and unpublished data available from the U.S. Department of Education.

Note: Details may not add to totals due to rounding.

[a.] Under the Emergency Impact Aid program, $1.9 million of the $880 million appropriated was not allocated to states, as the funds were not needed. Thus, the total appropriated amount is higher than the amount allocated and shown on the table.

Waiver of Vocational Rehabilitation State Grant Matches

The Assistance to Individuals with Disabilities Affected by Hurricane Katrina or Rita Act of 2005 (P.L. 109-82) provided the Secretary with the authority to waive the nonfederal match requirements for disaster-affected states. The law also gave disaster-affected states preference in claiming VR State Grant funds that were made available for reallocation if a state did not meet its full nonfederal match. The law defined an affected state as "a state that contains an area, or that received a significant number of individuals who resided in an area, in which the President has declared that a major disaster exists."

ACKNOWLEDGMENTS

This chapter draws largely on a previous report authored by Cassandria Dortch, Rebecca R. Skinner, and David P. Smole.

In: Emergency Management
Editor: Stephanie Padbury
ISBN: 978-1-53614-101-6

Chapter 5

FEMA Individual Assistance Programs: In Brief*

Shawn Reese
Emergency Management and Homeland Security Policy

Abstract

When the President declares a major disaster pursuant to the Robert T. Stafford Disaster Relief and Emergency Assistance Act (P.L. 93-288), the Federal Emergency Management Agency (FEMA) advises the President about types of federal assistance administered by FEMA available to disaster victims, states, localities, and tribes. The primary types of assistance provided under a major disaster declaration include funding through the Public Assistance program, Mitigation Assistance programs, and the Individual Assistance program.

The Public Assistance program provides federal financial assistance to repair and rebuild damaged facilities and infrastructure. Mitigation Assistance programs provide funding for jurisdictions, states, and tribes to ensure damaged facilities and infrastructure are rebuilt and reinforced to better withstand future disaster damage. Finally, the Individual

---

* This is an edited, augmented and reformatted version of a Congressional Research Service report, R45085, dated January 31, 2018.

Assistance program provides funding for basic needs for individuals and households following a disaster.

Eligible activities under the Individual Assistance program include funding for such things as mass care, crisis counseling, and temporary housing. FEMA advises the President on the type of individual assistance to be granted following each disaster, and works with state and local authorities in determining what assistance programs would best suit the needs within the disaster area. FEMA makes this determination based on a list of criteria designed to align federal disaster assistance with unmet needs in disaster-impacted areas.

This chapter provides a short summary of the types of individual assistance programs administered by FEMA following a disaster. This chapter also provides a summary of the criteria FEMA uses in determining which individual assistance programs may be made available to impacted areas following a major disaster declaration, and discusses a proposed rule to change these criteria.

INTRODUCTION

When the President declares a major disaster under the Robert T. Stafford Disaster Relief and Emergency Assistance Act,[1] the disaster declaration designates the types of Federal Emergency Management Agency (FEMA) assistance to be provided.[2] There are three primary types of assistance: Public Assistance (PA),[3] which addresses repairs to a community and state or tribal government infrastructure; Mitigation Assistance (MA), which provides funding for projects a state or tribe proposes to reduce the threat of future damage; and Individual Assistance (IA), which provides help to individuals and households affected by a major disaster.

While all major disaster declarations include MA, FEMA uses information on a range of "factors" from the Preliminary Damage Assessment (PDA) to determine whether PA and/or IA should be

[1] 42 U.S.C. §5170, as amended.

[2] For more information on Stafford Act declarations, see CRS Report R43784, *FEMA's Disaster Declaration Process: A Primer*, by Bruce R. Lindsay.

[3] For more information on PA, see CRS Report R43990, *FEMA's Public Assistance Grant Program: Background and Considerations for Congress*, by Jared T. Brown and Daniel J. Richardson.

recommended.[4] Following Hurricane Sandy, Congress required FEMA to revise and update the factors it considers when making a recommendation to the President regarding whether a major disaster declaration should include IA or not.[5] This chapter provides information on FEMA's IA programs and the factors FEMA uses to determine if IA should be part of a disaster declaration.

INDIVIDUAL ASSISTANCE PROGRAMS

FEMA considers numerous factors (detailed later in this chapter) when it recommends IA to the President following a major disaster. IA can include several programs, depending on whether the governor of the affected state or the tribal leader requests that specific type of FEMA assistance. FEMA's IA includes (1) Mass Care and Emergency Assistance, (2) Crisis Counseling Assistance and Training Program, (3) Disaster Unemployment Assistance, (4) Disaster Legal Services, (5) Disaster Case Management, and (6) the Individuals and Households Program.[6]

1) *Mass Care* includes directly supporting congregate sheltering, feeding, and hydration; distributing emergency supplies; and reuniting children with their parents/legal guardians, as well as adults with their families. *Emergency Assistance* encompasses a variety of services and functions, including coordination of volunteer organizations and unsolicited donations, managing unaffiliated volunteers and community relief services, supporting transitional sheltering, and supporting mass evacuations.

---

[4] In general, when a request for a major disaster declaration is submitted, representatives from FEMA meet with the state or tribal government and compile a PDA. FEMA then makes a recommendation to the President concerning whether a declaration should be issued. The President has the authority to make the declaration or deny the request. For more information on the PDA process, see CRS Report R44977, *Preliminary Damage Assessments for Major Disasters: Overview, Analysis, and Policy Observations*, coordinated by Bruce R. Lindsay.

[5] P.L. 113-2, Division B.

[6] P.L. 93-288, Title IV, §401-426.

2) The *Crisis Counseling Assistance and Training Program* assists individuals and communities recovering from the effects of a disaster through community-based outreach and psycho-educational services.[7] The program supports short-term counseling of disaster survivors. The program also provides information on coping strategies and emotional support by linking survivors with other individuals and agencies that help them in the recovery process.
3) *Disaster Unemployment Assistance* provides information and resources to individuals who were employed or self-employed, or were scheduled to begin employment during a disaster. It may also be provided to those who can no longer work or perform their job duties due to damage to their place of employment, do not qualify for regular unemployment benefits from a state, or cannot perform work or self-employment due to an injury as a direct result of a disaster.[8]
4) *Disaster Legal Services* provides legal assistance to low-income individuals who are unable to secure adequate legal services that meet their disaster-related needs.
5) *Disaster Case Management* provides a partnership between a case manager and the disaster survivor to assist them in carrying out a disaster recovery plan. The recovery plan includes resources, services, decisionmaking priorities, progress reports, and the goals needed to close their case.
6) *Individuals and Households Program* is comprised of two categories of assistance: Housing Assistance and Other Needs Assistance (ONA).
   a) Housing Assistance may include financial assistance to
      - reimburse for hotels, motels, or other short-term lodging;

---

[7] Psycho-educational services consist of therapeutic treatment for disaster victims that provides information and support to help them better understand and cope with their situation.

[8] For more information on disaster unemployment assistance, see CRS Report RS22022, *Disaster Unemployment Assistance (DUA)*, by Julie M. Whittaker.

- rent alternate housing accommodations while the applicant is displaced from their primary residence;
- repair a primary residence;
- assist in replacing owner-occupied residences when the residence is destroyed; and
- enter into lease agreements with owners of multifamily rental properties located in the disaster area.[9]

b) Housing Assistance may also include home repair and construction services provided in insular areas outside the continental United States and other locations where no alternative housing resources are available and where types of FEMA housing assistance that are normally provided (such as rental assistance) are unavailable, infeasible, or not cost-effective.

c) FEMA may provide manufactured housing units as a form of temporary housing through its Transitional Sheltering Assistance program.

d) Other Needs Assistance provides financial assistance for other disaster-related expenses and needs. These include

- child care;
- medical and dental expenses;
- funeral and burial costs; and
- transportation.

IA accounted for approximately 14.4% of all projected federal spending ($9.1 billion of $63.2 billion) from the Disaster Relief Fund for Stafford Act declarations occurring from FY2007 through FY2016, according to FEMA data as of August 10, 2017. Some types of FEMA IA are subject to cost sharing, such as ONA, which is subject to a cost share between FEMA and the state, territorial, or tribal government. FEMA covers 75% of eligible ONA costs, and the state, territorial, or tribal

---

[9] Housing assistance is capped at $33,300 per household.

government is responsible for the remaining 25%.[10] Transitional Sheltering under Mass Care and Emergency Assistance is also subject to a 75/25 percent cost share.

IA Factors for a Major Disaster Declaration

The factors FEMA uses to determine potential IA for disaster survivors have not been changed since originally published in 1999. The factors include

- *concentration of damages*—FEMA evaluates the concentrations of damages to individuals, and high concentrations of damages generally indicate a greater need for federal assistance than widespread or scattered damages throughout an area, region, or state;
- *trauma*—FEMA considers the degree of trauma to a state and communities, and some of the conditions that might cause such trauma include
  - large numbers of injuries and death,
  - large-scale disruption of normal community functions and services, and
  - emergency needs such as extended or widespread loss of power or water;
- *special populations*—FEMA considers whether special populations, such as low-income, the elderly, or the unemployed are affected, and whether they may have a greater need for assistance;
- *voluntary agency assistance*—FEMA considers the extent to which voluntary agencies and state or local programs can meet the needs of the disaster victims;

[10] 42 U.S.C. §5174(g) and 44 C.F.R. §206.110(i).

- *insurance*—FEMA considers the amount of insurance coverage in the disaster area because, by law,[11] federal disaster assistance cannot duplicate insurance coverage; and
- *average amount of individual assistance by state*—FEMA's regulations state that "[t]here is no set threshold for recommending Individual Assistance, but the following averages [see Table 1] may prove useful to States and voluntary agencies as they develop plans and programs to meet the needs of disaster victims."[12]

When reviewing disaster declarations between January 2008 and July 2013, FEMA found that requests with more than $7.5 million in assistance usually resulted in a declaration for IA. However, FEMA has not suggested using that amount as a threshold for assistance. Instead, it is providing several factors that relate directly to the Individual and Household Program (IHP), which is the primary IA program. In addition, other factors may provide greater detail for decisionmakers for IHP and for other prominent IA programs such as Crisis Counseling and Disaster Unemployment Assistance (DUA).

In response to congressional direction[13] for updated IA factors, FEMA issued a Notice of Proposed Rulemaking (NPRM) on November 12, 2015. FEMA published the proposed rule on November 12, 2016, and was accepting comments until January 11, 2016. FEMA has yet to finalize the proposed rule. The new proposed factors are

- state fiscal capacity and resource availability;
- uninsured home and personal property losses;
- disaster-impacted population profile;
- impact to community infrastructure;
- casualties; and
- disaster-related unemployment.

---

[11] 42 U.S.C. §5155.

[12] 44 C.F.R. §206.48(b).

[13] P.L. 113-2, Division B.

Table 1. Average Amount of Assistance per Disaster July 1994-July 1999

	Small states <2M pop.	**Medium states 2-10M pop.**	**Large states >10M pop.**
Average population (1990 census data)	*1.0M*	*4.7M*	*15.5M*
Disaster Housing Applications Approved	*1,507*	*2,747*	*4,679*
Homes Estimated Major Damage/Destroyed	*173*	*582*	*801*
Housing Assistance	$2.8M	$4.6M	$9.5M
Individual and Family Grant (IFG) Applications Approved	*495*	*1,377*	*2,071*
Individual and Family Grant Assistance	$1.1M	$2.9M	$4.6M
Disaster Housing/IFG Combined Assistance	$3.9M	$7.5M	$14.1M

Source: Data from FEMA's National Processing Service Centers, as presented in 44 C.F.R. §206.48(b).
Note: Data used are only from July 1994 to July 1999. M = millions.

FEMA proposed using three factors to evaluate a state and local jurisdiction's fiscal capacity: the total tax revenue of the state (a measurement recommended by the U.S. Government Accountability Office [GAO]),[14] the Gross Domestic Product (GDP) by state, and the per-capita personal income by local area. Using those factors, along with the projected IA costs, FEMA would then develop Individual Assistance Cost to Capacity (ICC) ratios.

As mentioned earlier, the estimates of potential FEMA assistance costs are generally derived from Preliminary Damage Assessments (PDAs) that are conducted after an event to help both the state and federal governments evaluate the situation. Using figures from this process for 153 major disaster declaration requests between January of 2008 and July of 2013, FEMA found that generally, the higher the ICC ratio, the more likely that IA was part of a major disaster declaration.

---

[14] See U.S. General Accounting Office, *Disaster Assistance: Improvement Needed in Disaster Declaration Criteria and Eligibility Assurance Procedures*, GAO-01-837, August 31, 2001, available at http://www.gao.gov/assets/240/ 232622.pdf.

FEMA emphasized that the proposed approach would not be a hard "threshold," nor did it anticipate incorporating it into its regulations because one single factor would not determine each decision on recommendations to the President.

A hard "threshold" was a recommendation of some FEMA stakeholders. It is likely they wanted to see a number comparable to the per capita amount employed as a factor for PA declarations. However, in response to that request, FEMA invoked Section 320 of the Stafford Act and suggested it chose not to violate the principles of Section 320. That section forbids a denial of assistance based *solely* (emphasis added) on an "arithmetic formula." While some formulas have been routinely used by FEMA in the process of making recommendations, according to FEMA, formulas have only been one factor among several considered. Instead, FEMA indicates its desire that such formulas, when used as a part of the PDA process, would influence a state's decision on whether to make a request, but would not remove the state's discretion.

FEMA's proposal offered more detail than previously available regarding state capacity, the tools FEMA may employ to consider individual state resources, and FEMA's own approach to assessing that capacity. But the proposed ICC and other factors raise questions as to how various factors may or may not be weighted in importance when considering a governor's request.

In addition, the proposed factors appeared to relate directly to current IA programs such as Disaster Unemployment Assistance and Crisis Counseling. The proposed rule also appeared to encourage the development of state programs that might replicate or complement FEMA-style assistance to families and individuals at the state level. It could be argued that such programs at the state level could be of assistance when federal supplemental help is not contemplated. At this time there is no information on when FEMA intends to finalize this proposed rule on revising IA factors.

In: Emergency Management
Editor: Stephanie Padbury
ISBN: 978-1-53614-101-6

Chapter 6

UNITED STATES FIRE ADMINISTRATION: AN OVERVIEW*

Lennard G. Kruger
Science and Technology Policy

ABSTRACT

The United States Fire Administration (USFA)—which includes the National Fire Academy (NFA)—is currently housed within the Federal Emergency Management Agency (FEMA) of the Department of Homeland Security (DHS). The objective of the USFA is to significantly reduce the nation's loss of life from fire, while also achieving a reduction in property loss and nonfatal injury due to fire.

The Consolidated Appropriations Act, 2018 (P.L. 115-141) provided $44.397 million for USFA. This total included $1.497 million for the National Emergency Training Center. For FY2019, the Administration requested $44.993 million for USFA. The FY2019 level consisted of $43.493 million under Education, Training, and Exercises in the Federal Assistance account, and $1.5 million for annual capital improvement of the National Emergency Training Center under the Procurement, Construction, and Improvements account.

---

* This is an edited, augmented and reformatted version of a Congressional Research Service report, RS20071, dated March 27, 2018.

On January 3, 2018, the President signed the United States Fire Administration, AFG, and SAFER Program Reauthorization Act of 2017 (P.L. 115-98). P.L. 115-98 extends the USFA authorization through FY2023. The authorization levels are the same as in the previous authorization: $76,490,890 each year for FY2017 through FY2023.

Meanwhile, concerns over the federal budget deficit could impact future funding levels for the USFA. Debate over the USFA budget has focused on whether the USFA is receiving an appropriate level of funding to accomplish its mission, given that appropriations for USFA have consistently been well below the agency's authorized level. Additionally, an ongoing issue is the viability and status of the USFA and the National Fire Academy within the Department of Homeland Security.

BACKGROUND

The United States Fire Administration (USFA) is currently an entity within the Federal Emergency Management Agency (FEMA) of the Department of Homeland Security (DHS). Its mission is to provide leadership, coordination, and support for the nation's fire prevention and control, fire training and education, and emergency medical services activities, and to prepare first responders and health care leaders to react to hazard and terrorism emergencies of all kinds. One of USFA's key objectives is to significantly reduce the nation's loss of life from fire, while also achieving a reduction in property loss and nonfatal injury due to fire. Although fire loss has improved significantly over the past 25 years, the fire problem in the United States remains serious. The United States still has one of the highest fire death rates in the industrialized world. According to the National Fire Protection Association (NFPA), in 2015 there were 1,345,500 total fires reported, 3,280 civilian fire deaths, 15,700 civilian fire injuries, and an estimated $14.3 billion in direct property loss.[1] There were 69 on-duty firefighter deaths in 2016.[2]

---

[1] National Fire Protection Association, *Fire Loss in the United States During 2015*, September 2016, available at http://www.nfpa.org/news-and-research/fire-statistics-and-reports/fire-statistics/fires-in-the-us/overall-fire-problem/ fire-loss-in-the-united-states.

[2] National Fire Protection Association, *Firefighter Fatalities in the United States—2016*, June 2017, available at http://www.nfpa.org/news-and-research/fire-statistics-and-reports/fire-statistics/the-fire-service/fatalities-and-injuries/ firefighter-fatalities-in-the-united-states.

The genesis of USFA and FEMA's fire prevention and control activities can be found in the landmark 1973 report of the National Commission on Fire Prevention and Control,[3] entitled *America Burning*. The commission recommended the creation of a federal fire agency which would provide support to state and local governments and private fire organizations in their efforts to reduce fire deaths, injuries, and property loss. The commission recommended that this new agency be placed within the Department of Housing and Urban Development. Congress instead opted to place the agency in the Department of Commerce, and with the passage of the Federal Fire Prevention and Control Act of 1974 (P.L. 93-498),[4] the National Fire Prevention and Control Administration (NFPCA) was established. In 1978, Congress changed the name of NFPCA to USFA (P.L. 95-422), and in 1979, President Carter's Reorganization Plan No. 3 placed the USFA within the newly created FEMA. Also in 1979, the National Fire Academy (NFA) in Emmitsburg, MD, was opened, offering courses and training to fire service personnel and other persons engaged in fire prevention and control.

During the early 1980s, the Reagan Administration proposed the elimination of the USFA (while preserving the Fire Academy). Although Congress did not allow the termination of the USFA, the agency suffered severe staff reductions and the Fire Academy was separated from the USFA and housed organizationally with other FEMA emergency training programs. In 1991, the NFA was subsequently reorganized back into the USFA, where it remains today.

Currently, the USFA is located on the grounds of the National Emergency Training Center in Emmitsburg, MD. USFA programs include the following:

Data Collection

USFA's National Fire Data Center (NFDC) administers a national system (the National Fire Incident Reporting System or NFIRS) used for collecting, analyzing, and disseminating data and information on fire and

---

[3] The commission was created by P.L. 90-259, the Fire Research and Safety Act of 1968.

[4] 15 U.S.C. 2201 et seq.

other emergency incidents to state and local governments and the fire community. The NFDC provides a national analysis of the fire problem, identifying problem areas for which prevention and mitigation strategies are needed.

Public Education and Awareness

Through partnerships and special initiatives, USFA involves the fire service, the media, other federal agencies, and safety interest groups in the development and delivery of fire safety awareness and education programs. These programs are targeted at those groups most vulnerable to the hazards of fire, including the young, elderly, and disabled.

Training

USFA's National Fire Academy (NFA) offers educational opportunities for the advanced professional development of the mid-level and senior fire/EMS officers and allied professionals involved in fire prevention and life safety activities. The academy develops and delivers educational and training programs with a national focus that supplement and support state and local fire service training. The NFA also offers training to support the National Incident Management System Integration Center (NIC) and nationwide implementation of the National Incident Management System (NIMS).

Research and Technology

Through research, testing, and evaluation, USFA works with public and private entities to promote and improve fire and life safety. Research and special studies are conducted on fire detection, suppression, and notification systems, as well as issues related to firefighter and emergency responder health and safety. Research results are published and made available to the public free of charge through the USFA Publications Center.

In fulfilling its mission, the USFA uses the assets of the National Fire Academy, the National Emergency Training Center (NETC) Facilities and Support Services, and the National Fire Programs Division.

On May 18, 2017, President Trump announced his intention to appoint Chief G. Keith Bryant as the USFA Administrator.[5] G. Keith Bryant was sworn in as the U.S. Fire Administrator on August 4, 2017.

BUDGET

The USFA receives its annual appropriation through the House and Senate Appropriations Subcommittees on Homeland Security. Table 1 shows recent and proposed appropriated funding for USFA.

Appropriations

Beginning in FY2004, the USFA was funded through the Preparedness, Mitigation, Response, and Recovery (PMRR) account within the Emergency Preparedness and Response Directorate of the Department of Homeland Security. On July 13, 2005, then-DHS Secretary Michael Chertoff announced a restructuring of DHS, effective October 1, 2005. USFA was removed from the PMRR account and received a separate appropriation (its own line item) under the new DHS Directorate for Preparedness. The FY2007 Department of Homeland Security appropriations bill (P.L. 109-295) transferred the USFA back to the Federal Emergency Management Agency within DHS.

FY2016

The Obama Administration's FY2016 budget proposed $41.582 million for USFA, a 5.6% decrease from the FY2015 level. Of the requested total appropriation, $12.268 million would be allocated to the National Fire Academy, $10.981 million to National Fire Programs, and

[5] The White House, Office of the Press Secretary, "President Donald J. Trump Announces Key Additions to His Administration," May 18, 2017, available at https://www.whitehouse.gov/the-press-office/2017/05/18/president-donald-j-trump-announces-key-additions-his-administration.

$18.333 million to National Emergency Training Center (NETC) Management, Operations, and Support.

Table 1. Appropriations, U.S. Fire Administration (millions of dollars)

	FY2017 (P.L. 115-31)	FY2018 (Admin. request)	FY2018 (P.L. 115-141)	FY2019 (Admin. request)
USFA	44[a]	43.4[b]	44.4[c]	45[d]

[a] Includes $42.5 million for USFA under the Federal Assistance account and $1.5 million under the Procurement, Construction, and Improvements account.

[b] Includes $41.9 million for USFA under the Federal Assistance account and $1.5 million under the Procurement, Construction, and Improvements account.

[c] Includes $42.9 million for USFA under the Federal Assistance account and $1.5 million under the Procurement, Construction, and Improvements account.

[d] Includes $43.5 million for USFA under the Federal Assistance account and $1.5 million under the Procurement, Construction, and Improvements account.

On June 18, 2015, the Senate Appropriations Committee reported S. 1619, the Department of Homeland Security Act, 2016. The Senate bill would provide $44 million for USFA, the same level that was provided in FY2015. In the accompanying report (S.Rept. 114-264), the committee encouraged FEMA to use the amount over the request to continue development of NFIRS. The committee expressed concern over the NFIRS and recommended that FEMA address the shortfalls and long-term planning needs of NFIRS in future budget submissions. The committee also directed USFA to continue its traditional funding for the National Fallen Firefighters Memorial and related activities.

On July 21, 2015, the House Appropriations Committee approved H.R. 3128, the Department of Homeland Security Appropriations Act, 2016. The bill would provide $44 million for USFA, the same as the amount provided in FY2015.

On December 18, 2015, the President signed the Consolidated Appropriations Act, 2016 (P.L. 114-113). The law provided $44 million for USFA in FY2016.

FY2017

The Administration's FY2017 budget proposed \$42.3 million for USFA, a 3.8% decrease from the FY2016 level. The request included \$1.5 million for facilities improvement under the Procurement, Construction, and Improvements account. The budget proposal included \$500,000 for distance learning capability and reductions of \$1 million each for NFIRS and state fire training grants. The budget request would also transfer the stand-alone USFA budget account into the Preparedness and Protection activity under FEMA's broader Federal Assistance account.

On May 26, 2016, the Senate Appropriations Committee approved S. 3001, the Department of Homeland Security Act, 2017. The Senate bill would provide \$44 million for USFA, which matches the FY2016 level and is \$1.688 million above the request. In the accompanying report (S.Rept. 114-68), the committee stated that the increase over the Administration request should allow for the continued development of NFIRS and support for the National Fallen Firefighters Memorial. The committee maintained a separate budget account for USFA and did not transfer the USFA budget account to the Federal Assistance account as proposed in the Administration budget request.

On June 22, 2016, the House Appropriations Committee approved its version of the Department of Homeland Security Appropriations Act, 2017. Unlike the Senate, the House committee transferred the USFA budget account into a broader "Federal Assistance" account in FEMA. The bill provided \$42.5 million for USFA under the Federal Assistance account and \$1.5 million under Procurement, Construction, and Improvements for National Fire Academy facility costs.

The Consolidated Appropriations Act, 2017 (P.L. 115-31) funded USFA at a total level of \$44 million in FY2017. This consisted of \$42.5 million under Education, Training, and Exercises in the Federal Assistance account, and \$1.5 million under the Procurement, Construction, and Improvements account.

FY2018

For FY2018, the Administration requested $43.41 million for USFA, slightly below the FY2017 level of $44 million. The FY2018 level consists of $41.913 million under Education, Training, and Exercises in the Federal Assistance account, and $1.497 million under the Procurement, Construction, and Improvements account. According to the FY2018 budget proposal, the request reflects a $1 million reduction to the State Fire Training Assistance grants.

On July 18, 2017, the House Appropriations Committee approved the Department of Homeland Security Appropriations Act, 2018 (H.R. 3355; H.Rept. 115-239). The bill provided the same level as the Administration request: $41.913 million under Education, Training, and Exercises in the Federal Assistance account, and $1.497 million under the Procurement, Construction, and Improvements account.

On September 14, 2017, the House passed H.R. 3354, a FY2018 omnibus appropriations bill that includes funding for USFA. During floor consideration, the House adopted an amendment offered by Representative Pascrell that added $1 million for USFA's State Fire Training Assistance grants, thereby restoring the Administration's proposed reduction. H.R. 3354 would provide a total of $44.41 million for USFA.

The Consolidated Appropriations Act, 2018 (P.L. 115-141) provided $44.397 million for USFA. This total included $1.497 million in the FEMA Procurement, Construction, and Improvements account for the National Emergency Training Center. State Fire Training Assistance grants continued to be funded by USFA.

FY2019

For FY2019, the Administration requested $44.993 million for USFA. The FY2019 level consisted of $43.493 million under Education, Training, and Exercises in the Federal Assistance account, and $1.5 million for annual capital improvement of the National Emergency Training Center under the Procurement, Construction, and Improvements account.

Authorizations

The U.S. Fire Administration Reauthorization Act of 2003 (P.L. 108-169) was signed into law on December 6, 2003. The act reauthorized the USFA through FY2008 at the following levels: $63 million for FY2005, $64.85 million for FY2006, $66.796 million for FY2007, and $68.8 million for FY2008. P.L. 108-169 also reestablished the presidentially appointed position of the U.S. Fire Administrator, which had been statutorily abolished by the Homeland Security Act of 2002. Additionally, the legislation directed the USFA to develop new firefighting technologies and standards in coordination with private sector standards groups and federal, state, and local agencies. P.L. 108-169 required that equipment purchased with fire grant money meet or exceed voluntary consensus standards when feasible.

The U.S. Fire Administration Reauthorization Act of 2008 was signed into law on October 8, 2008 (P.L. 110-376). P.L. 110-376 authorized the USFA at $70 million for FY2009, $72.1 million for FY2010, $74.263 million for FY2011, and $76.491 million for FY2012. Provisions included authorizing National Fire Academy training program modifications and reports; directing the National Fire Academy to provide training on incidents occurring in the wildfire-urban interface, multi-jurisdictional fires, hazardous materials incidents, and advanced emergency medical services; authorizing USFA to enter into contracts with one or more nationally recognized third-party organizations to deliver training; a report on the feasibility of providing incident command training for fires at ports and in marine environments; national fire incident reporting system upgrades; sponsoring and disseminating research on fire prevention and control at the wildland-urban interface; encouraging adoption of national voluntary consensus standards for firefighter health and safety; establishing a state and local fire service position at the National Operations Center within DHS; providing coordination regarding fire prevention and control and emergency medical services; and expressing congressional support for USFA recommendations for adoption and education regarding sprinklers in commercial and residential buildings.

On January 2, 2013, the President signed P.L. 112-239, the FY2013 National Defense Authorization Act. Title XVIII, Subtitle B was the U.S. Fire Administration Reauthorization Act of 2012, which authorized USFA through FY2017. P.L. 112-239 included the following provisions:

- reauthorized USFA at an annual level of $76,490,890 for FY2013 through FY2017, and for each fiscal year sets aside $2,753,672 to be used to carry out Section 8(f) of the Fire Prevention and Control Act (15 U.S.C. 2207) related to evaluation of technology and development of standards;
- authorized the USFA Administrator to appoint a Deputy Administrator;
- authorized the Administrator to take such steps as the Administrator considers appropriate to educate the public and overcome public indifference as to fire, fire prevention, and individual preparedness; and
- removed the limitation on funding levels for updating the National Fire Incident Reporting System.

In the 115th Congress, on July 12, 2017, the House Subcommittee on Research and Technology, Committee on Science, Space and Technology, held a hearing entitled *U.S. Fire Administration and Fire Grant Programs Reauthorization: Examining Effectiveness and Priorities*. Testimony was heard from the USFA acting administrator and from fire service organizations.[6]

On December 15, 2017, H.R. 4661, the United States Fire Administration, AFG, and SAFER Program Reauthorization Act of 2017, was introduced by Representative Comstock, which sought to reauthorize the USFA through FY2023. On December 18, 2017, the House passed H.R. 4661 by voice vote under suspension of the rules. On December 21, 2017, the Senate passed H.R. 4661 without amendment by unanimous consent.

---

[6] Testimony is available at https://science.house.gov/legislation/hearings/subcommittee-research-and-technology-hearing-us-fire-administration-and-fire.

On January 3, 2018, the President signed the United States Fire Administration, AFG, and SAFER Program Reauthorization Act of 2017 (P.L. 115-98). P.L. 115-98 extends the USFA authorization through FY2023. The authorization levels are the same as in the previous authorization: $76,490,890 each year for FY2017 through FY2023, of which $2,753,672 each fiscal year is to be used to carry out Section 8(f) of the Fire Prevention and Control Act (15 U.S.C. 2207) related to evaluation of technology and development of standards.

ASSISTANCE TO FIREFIGHTERS PROGRAM (FIRE ACT GRANTS)

The Assistance to Firefighters Grant (AFG) Program,[7] also known as the FIRE Act grant program, was established by Title XVII of the FY2001 Floyd D. Spence National Defense Authorization Act (P.L. 106-398). The program provides federal grants directly to local fire departments and unaffiliated Emergency Medical Services (EMS) organizations to help address a variety of equipment, training, and other firefighter-related and EMS needs. A related program is the Staffing for Adequate Fire and Emergency Response Firefighters (SAFER) program, which provides grants for hiring, recruiting, and retaining firefighters.[8]

Since its inception, the fire grant program has been administered by FEMA/USFA (FY2001- FY2003), the Office for Domestic Preparedness (FY2004), the Office of State and Local Government Coordination Preparedness (FY2005), and the Office of Grants and Training in the DHS Directorate for Preparedness (FY2006). The FY2007 DHS Appropriations Act (P.L. 109- 295) transferred USFA to FEMA and the fire and SAFER grants to the Grants Programs Directorate in FEMA. Congressional appropriations reports have consistently instructed DHS to maintain USFA

---

[7] For more information on the fire grant program, see CRS Report RL32341, *Assistance to Firefighters Program: Distribution of Fire Grant Funding*, by Lennard G. Kruger.

[8] For more details on the SAFER grant program, see CRS Report RL33375, *Staffing for Adequate Fire and Emergency Response: The SAFER Grant Program*, by Lennard G. Kruger.

involvement in the grant administration process for AFG and SAFER grants.

In September 2016, the Government Accountability Office (GAO) released a report entitled *Fire Grants: FEMA Could Enhance Program Administration and Performance Assessment.* Among its findings, GAO concluded that FEMA has not defined and documented USFA's specific role or responsibilities with the fire grants program, and that there is no formalized relationship or policy regarding how the two organizations' programs could work together. According to GAO

> Although a level of informal coordination exists between GPD [Grant Programs Directorate] and USFA, enhancing these efforts by using collaborative mechanisms that our work across the federal government has identified as key features and issues to consider during implementation—such as clearly defining and agreeing upon USFA's role and responsibilities and documenting agreement regarding how they will be collaborating—could help GPD further leverage USFA expertise and resources in support of the fire grants programs, which could also help GPD manage the integration of fire grants into broader national preparedness efforts.[9]

In December 2016, the USFA signed an agreement with FEMA's Grant Programs Directorate to provide a framework for each entity's roles and responsibilities for improving the management of the fire grants.[10]

---

[9] Government Accountability Office, *Fire Grants: FEMA Could Enhance Program Administration and Performance Assessment*, GAO-16-744, September 2016, p. 31, available at http://gao.gov/assets/680/679787.pdf.

[10] Testimony of Denis Onieal, Acting U.S. Fire Administrator, before the House Committee on Science, Space and Technology, Subcommittee on Research and Technology, July 12, 2017, available at https://science.house.gov/sites/republicans.science.house.gov/files/ documents/ HHRG-115-SY15-WState-DOnieal-20170712.pdf.

ISSUES IN THE 115TH CONGRESS

Concerns over the federal budget deficit could impact future funding levels for the USFA. Debate over the USFA budget has focused on whether the USFA is receiving an appropriate level of funding to accomplish its mission, given that appropriations for USFA have consistently been well below the agency's authorized level, and given that USFA's budget has remained flat over recent years.

An ongoing issue is the viability and status of the USFA and National Fire Academy within the Department of Homeland Security. While supportive of the reorganization of FEMA into DHS, many in the fire service community have cautioned that USFA and NFA programs—which address the day-to-day challenges faced by fire departments—should not be overshadowed in an organization which focuses on homeland security and counterterrorism. Since the establishment of DHS in March 2003, fire service groups have opposed a number of actions DHS has taken with respect to the USFA and NFA. These included the abolishment of the presidentially appointed position of U.S. Fire Administrator (subsequently reestablished by enactment of the USFA Reauthorization Act of 2003); proposed cancellations of some NFA courses in 2003 due to an across-the-board FEMA budget cut (those NFA courses were subsequently restored after fire service protests); and the transfer of the fire grant program from the USFA to the DHS Office for Domestic Preparedness.

INDEX

D

E

F

O

P

R

S

T

U

V

W

Y